PALMPRINT AUTHENTICATION

Kluwer International Series on Biometrics

Professor David D. Zhang　　**Professor Anil K. Jain**
Consulting Editors

Department of Computer Science　　*Dept. of Computer Science & Eng.*
Hong Kong Polytechnic University　　*Michigan State University*
Hung Hom, Kowloon, Hong Kong　　*3115 Engineering Bldg.*
　　East Lansing, MI 48824-1226, U.S.A.

Email: csdzhang@comp.polyu.edu.hk　　Email: jain@cse.msu.edu

In our international and interconnected information society, there is an ever-growing need to authenticate and identify individuals. Biometrics-based authentication is emerging as the most reliable solution. Currently, there have been various biometric technologies and systems for authentication, which are either widely used or under development. The Kluwer International Book Series on Biometrics (KISB) will systematically introduce these relative technologies and systems, presented by biometric experts to summarize their successful experience, and explore how to design the corresponding systems with in-depth discussion.

In addition, this series aims to provide an international exchange for researchers, professionals, and industrial practitioners to share their knowledge of how to surf this tidal wave of information. KISB will contain new material that describes, in a unified way, the basic concepts, theories and characteristic features of integrating and formulating the different facets of biometrics, together with its recent developments and significant applications. Different biometric experts, from the global community, are invited to write these books. Each volume will provide exhaustive information on the development in that respective area. KISB will provide a balanced mixture of technology, systems and applications. A comprehensive bibliography on the related subjects will also be appended for the convenience of our readers.

Additional titles in the series:

FACIAL ANALYSIS FROM CONTINUOUS VIDEO WITH APPLICATIONS TO HUMAN-COMPUTER INTERFACE *by Antonio J. Colmenarez, Ziyou Xiong and Thomas S. Huang;* ISBN: 1-4020-7802-1

COMPUTATIONAL ALGORITHMS FOR FINGERPRINT RECOGNITION *by Bir Bhanu and Xuejun Tan;* ISBN: 1-4020-7651-7

Additional information about this series can be obtained from our website:
http//www.wkap.nl

CONTENTS

PART I   INTRODUCTION AND BACKGROUND

Chapter 1

Chapter 2

PART II   OFFLINE METHODOLOGIES

Chapter 3

PREFACE

In recent times, an increasing, worldwide effort has been devoted to the development of automatic personal identification systems that can be effective in a wide variety of security contexts. As one of the most powerful and reliable means of personal authentication, biometrics has been an area of particular interest. This interest has led to the extensive study of biometric technologies such as fingerprint and face recognition and the development of numerous algorithms, applications, and systems. Palmprints, in particular, have attracted a lot of interest. Of course, the study of palmprints is not entirely recent - they have been have been used in fortune telling for more than 3,000 years - but from the point of view of personal authentication it is a comparatively new biometric. Recent interest in palmprints is justified. Palmprints have a number of unique advantages: they are rich in features such as principal lines, wrinkles, and textures and these provide stable and distinctive information sufficient for separating an individual from a large population. Having worked on palmprints since 1996, our team certainly regards the palmprint as a very effective biometric. Our first technical paper, "Automated personal identification by palmprint" was published in 1998, and discussed a new approach to automated personal identification using palmprints. This was followed with even more extensive investigations into palmprint technology, and this research has now evolved to a point that researchers expect palmprint can be used to compensate for the weaknesses of other biometrics technologies.

Initially, our palmprint research focused on offline inked palmprint images. We built three offline palmprint databases and then when our online palmprint acquisition device was developed we shifted focus to the use of online inkless palmprint images, ultimately creating two online palmprint databases. Most of the experiments in this book use palmprint images from these databases. Since that time, a number of algorithms have been proposed for use in palmprint technology, including segmentation approaches, feature extraction methodologies, matching strategies and classification ideas. Both this explosion of interest and this diversity of approaches have been reflected in the wide range of recently published technical papers, technical papers with a range of different focuses including palmprint registration, palm line extraction, texture feature extraction, and transform-based palmprint feature extraction. Unfortunately, this wonderful variety of information is available in an inconvenient range of journals and conference proceedings. This book seeks to gather and present in a unified way current knowledge relevant to the basic concepts, definition and characteristic features of palmprint technology and demonstrates a palmprint identification system prototype. We hope thereby to provide readers with a concrete survey of the field in one volume. Selected chapters provide in-depth guides to specific algorithm designs and implementations.

This book provides a comprehensive introduction to palmprint technologies. It reveals automatic techniques for palmprint authentication, from the first approach based on offline palmprint images, to the current state-of-the-art algorithm using online palmprint images. It is suitable for different levels of readers: those who want to learn more about palmprint technology, and those who wish to understand, participate in, and/or develop a palmprint authentication system. We have tried to keep explanations elementary without sacrificing depth of coverage or mathematical rigor. But some algorithms need special knowledge on specific topics such as wavelets, Fourier transforms, fisher discriminant analysis, or Gabor analysis. The first part of this book introduces biometrics and palmprint technology. The second part introduces offline palmprint methods. The third part considers algorithms for online palmprint technologies. We discuss in considerable depth various algorithm design methodologies, and architectures, as well as at the different stages of implementation of a palmprint system including palmprint segmentation, feature extraction, matching, and classification. The last part of this book demonstrates a palmprint acquisition system and a palmprint identification system prototype and, emphasizing system design, discusses the system framework and components, from software algorithms to hardware devices.

This book is a comprehensive introduction to both theoretical and practical in issues in palmprint authentication. It would serve as a textbook or as a useful reference for graduate students and researchers in the fields of computer science, electrical engineering, systems science, and information technology. Researchers and practitioners in industry and R&D laboratories working security system design, biometrics, immigration, law enforcement, control, and pattern recognition would also find much of interest in this book.

To encourage more people to join us in our research into palmprint technology, we have made available part of our palmprint database for the public to download. Further details can be found on http://www.comp.polyu.edu.hk/~biometrics/.

David Zhang

March 2004

Biometrics Research Centre (UGC/CRC)
The Hong Kong Polytechnic University

ACKNOWLEDGEMENTS

My sincere thank goes to Professor Zhaoqi Bian of Tsinghua University, Beijing, China, for his advice throughout this research. I would like to thank my team members, Dr. Wei Shu, Dr. Jane You, Guangming Lu, Adams Kong, Michael Wong, Xiangqian Wu, Wenxin Li and Dr. Lei Zhang for their hard work and unstinting support. In fact, this book is the common result of their many contributions. I would also like to express my gratitude to my research fellows, Dr. Xiaoyuan Jing, Dr. Jian Yang, Dr. Jie Zhou, Dr. Changshui Zhang, and Dr. Ajay Kumar for their invaluable help and support. Thanks are also due to Martin Kyle, Laura Liu and Qingyun Dai, for their help in the preparation of this book and to Michael Wong for his technical support. The financial support of UGC/CRC Grants in Hong Kong and NFSC fund in China are of course also greatly appreciated. I take this opportunity to thank all the contributors for agreeing to write for the book. I owe a debt of thanks to Susan Legerstrom-Fife and Sharon Palleschi of Kluwer Academic Publishers, for their initiative in publishing this volume.

PART I

INTRODUCTION AND BACKGROUND

1 INTRODUCTION

This chapter first gives an overview of biometrics technology, and the need for a biometrics system to enhance the security. Then, the architecture of a biometric system is revealed in Section 1.2. Some basic concepts such as the operation modes and performance evaluation of a biometric system are also reported. After that, a discussion on the hand based biometrics is presented in Section 1.3. Section 1.4 tries to illustrate why we need a new biometric, the palmprint, and what the advantages are of using palmprints. Finally, the book perspective is given in Section 1.5.

1.1 Overview

Personal identification and verification both play a critical role in our society. Today more and more business activities and work practices are computerized. e-Commerce applications such as e-Banking or security applications such as building access demand fast, real time, and accurate personal identification. Traditional knowledge-based or token-based personal identification or verification systems are tedious, time-consuming, inefficient and expensive.

Knowledge-based approaches use "something that you know" (such as passwords and personal identification numbers [1]) for personal identification; token-based approaches, on the other hand, use "something that you have" (such as passports or credit cards) for the same purpose. Tokens (e.g. credit cards) are time consuming and expensive to replace. Passwords (e.g. for computer login and email account) are hard to remember. According to Gartner [2], a company spends about US$14 to US$28 on handling a password reset and about 19% of help desk calls are related to the password reset problem. This may suggest that the traditional knowledge-based password protection is unsatisfactory. Thorough discussions of the weaknesses of the password authentication methods are mentioned in [3]. Since these approaches are not based on any inherent attribute of an individual in the identification process, they are unable to differentiate between an authorized person and an impostor who fraudulently acquires the "token" or "knowledge" of the authorized person. These shortcomings have led to biometrics identification or verification systems becoming the focus of the research community in recent years: biometric systems based on iris, hand geometry and fingerprint [4-8] were developed in the past decade.

Biometrics involves the automatic identification of an individual based on his physiological or behavioural characteristics. The first commercial system, Identimat, was developed in 1970s, as part of a time clock at Shearson Hamill, a Wall Street investment firm [9]. It measured the shape of the hand and the lengths of the fingers. At the same time, fingerprint-based automatic checking systems were widely used in law enforcement by the FBI and by the US government departments. Advances in hardware such as faster processing power and greater memory capacity made biometrics more viable. Since the 1990s, iris, retina, face, voice, palmprint, signature and DNA technologies have joined the biometric family [1, 10].

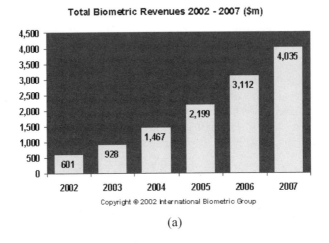

(a)

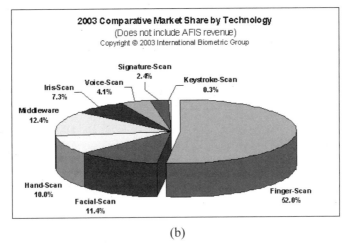

(b)

Figure 1-1. (a) Total biometric revenues prediction in 2002-2007, (b) Comparative market share by biometric technologies in 2003. (*Source: International Biometric Group*)

After the 911 terrorist attacks, the interest in biometrics-based security solutions and applications has increased dramatically, especially in the need to spot potential criminals in crowds. This further pushes the demand for the development of different biometrics products. For example, some airlines have implemented the iris recognition technology in airplane control rooms to prevent any entry by unauthorized persons. In 2004, all Australian international airports will implement passports using face recognition technology for airline crews and this will eventually becomes available to all Australian passport holders [11].

Figure 1-1 (a) shows predicted total revenues from biometrics for 2002-2007. A steady rise in revenues is predicted, from US $928 million in 2003 to US $4,035 million in 2007. Figure 1-1 (b) indicates the Comparative Market Share by different biometric technologies for the year 2003. With a 2003 market share of 52% [12], fingerprint technology is the world's most widespread biometric technology.

1.2   Biometric Systems

Biometric System Architecture

A biometric system is essentially a pattern recognition system which makes a personal identification by determining the authenticity of a specific physiological or behavioral characteristic possessed by the user [13]. Normally, personal characteristics such as fingerprints, palmprints or 3-D hand geometry are obtained through a sensor and fed into the pattern recognition engine to return a result of success or failure. Figure 1-2 shows the architecture of a typical biometric system. In general, biometric systems consist of the following four stages: 1) Data acquisition 2) Signal/Image preprocessing 3) Feature extraction and 4) Feature matching.

1) *Data acquisition* – Biometric data (signal/image) is obtained from an input device. The quality of signals is very important since they form the raw input for subsequent processing.
2) *Signal/Image preprocessing* – Enhancement of the signal/image is performed in this stage, including segmentation, noise reduction, and rotation and translation normalization.
3) *Feature extraction* – The features defined possess the stable and unique properties of low intra-class difference and high inter-class difference. These features are used to create a master template which is stored in the system database.
4) *Feature matching* – A matching score is obtained by matching the identification template against the master templates. If the score is less than a given threshold, the user is authenticated.

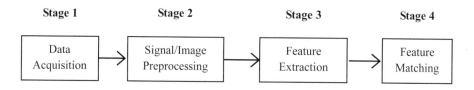

Figure 1-2. Four stages of a biometric system.

Operation Mode of a Biometric System

A biometric system is usually operated in three modes: enrollment, identification, and verification. Some systems, however, only have either identification or verification modes.

Enrollment – Before a user can be verified or identified by the system, he/she must be enrolled by the biometric system. The user's biometric data is captured, preprocessed and feature-extracted as shown in Stages 1-3 of Figure 1-2. The user's template is then stored in a database or file system.

Identification – This refers to the identification of a user based solely on his/her biometric information, without any prior knowledge about the identity of a user. Sometimes it is referred to as one-to-many matching or recognition. Identification applies the processes of stages 1-3 to create an identification template. The system will then retrieve all the templates from the database for feature matching in Stage 4. The match is either successful or not. Generally, accuracy decreases as the size of the database grows.

Verification – This requires that an identity (ID card, smart card or ID number) be claimed. A verification template is then matched with a master template to verify the person's identity claim. Sometimes verification is referred to as one-to-one matching or authentication.

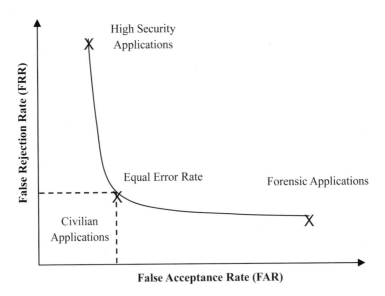

Figure 1-3. A hypothetical ROC curve and typical operating points for different biometric applications.

Performance Evaluation of a Biometric System

Performance evaluation of a biometric system usually refers to its False Acceptance Rate (FAR), False Rejection Rate (FRR) and Equal Error Rate (EER) [14, 33, 130]. FAR refers to a situation where a non-registered user gains access to a biometrically protected system, while FRR refers to a situation where a registered user fails to gain rightful access to a biometrically protected system at the first attempt. EER is represented by a percentage in which the FAR and FRR are equal. EER provides a

unique measurement which fairly compares the performance of differei
systems. In an ideal system, there are no false rejections and no false acce|
as yet no such system has been developed. In general, a biometric syste   _
operated at different levels of security: high, medium, or low. Figure 1-3 illustrates a
hypothetical Receiver Operating Characteristics (ROC) curve and typical operating
points for different biometric applications [15]. The ROC curve of a system
illustrates the false rejection rate (FRR) and the false acceptance rate (FAR) of a
matcher at all operating points (threshold, T). Each point on an ROC curve defines
the FRR and FAR operating at a particular threshold. The security threshold depends
mainly on the application context of a biometric system. If it is designed for a high
security environment such as the entrance to a nuclear facility, then more false
rejections with less false acceptances occur. The ideal case is to operate the system in
both low FRR and low FAR modes.

Different Biometric Technologies

A significant limitation on existing biometric-based personal identification systems is
their imperfect performance in terms of accuracy and acceptance. These systems
sometimes falsely accept an impostor (FAR) or falsely reject a genuine user (FRR)
[1]. An ideal biometric system that satisfies all the criteria of high accuracy, high user
acceptance and low cost is yet to be developed. Each existing system has its own
strengths and limitations. There can be no definite answer as to which biometric
system is the best as the selection of biometrics is an application-dependent decision.
Every system has its strengths and limitations. Iris recognition systems such as
IrisAccess™ from Iridian Technologies, Inc. [7] provide a very high level of
accuracy and security. Their scalability and fast processing power of iris system's
fulfils the rigid requirements of today's marketplace. However, they are expensive
and users find iris systems intrusive. 3D hand geometry system requires only small
feature size, including the length, width, thickness and surface area of the hand and
fingers of a user. In spite of its widespread use nowadays (10% market share in 2003
[12]), the uniqueness of the hand features is not guaranteed, making them less suited
to one-to-many identification applications. Fingerprint technology is the most widely
used biometric technology in the world. It had a market share of 52% in 2003 [12]. It
has been the premier online biometrics technology since the 1980s. It has advantages
of small chip size, easy to acquire and highly accurate, making it the most popular
biometric technology. However, some people may have fingerprints worn away due
to hand work and some old people may have many small creases around their fingers.
This lowers the system performance. In addition, some people may associate the
fingerprint acquisition process with the associated police process, making it less
acceptable to them. Compared to other biometrics, face verification is low cost,
needing only a camera mounted in a suitable position such as the entrance of a
physical access control area. For verification purposes it captures the physical
characteristics such as the upper outlines of the eye sockets, the areas around the
cheekbones, and the sides of the mouth. Face-scanning is suitable in environments
where screening and surveillance are required with minimal interference with
passengers. The State of Virginia in the United States has installed face recognition
cameras on Virginia's beaches to automatically record and compare faces with images
of suspected criminals and runaways [20]. However, the user acceptance of facial

scanning is lower than that of fingerprints, according to the IBG Report [21]. An off-the-shelf facial recognition system, FaceIt® ARGUS, detects and identifies human faces as they pass through a camera's field of view, which can help the detection of suspect from a database. The next section will focus on describing hand-based biometric technologies.

1.3   Hand-Based Biometrics

As mentioned in Section 1.2, the goal of automated biometric-based identification is to verify or recognize the identity of a live person on the basis of physiological or behavioral characteristics. There have been two current approaches, fingerprint identification and hand geometry identification, that are based upon the measurement and comparison of the features extracted from a person's hand [9, 17]. Hand geometry makes use of hand features for personal identification. A new biometric authenticates a person by using infrared light to obtain the vein patterns from a palm [29]. Each of these methods will now be discussed.

Fingerprint
The commercial application of biometric devices began in the early 1970s when a system called Identimat, which measured the shape of the hand and length of the fingers, was used as part of a time clock at Shearson Hamill, a Wall Street investment firm [9]. Subsequently, hundreds of Identimat devices were used to establish identity for physical access at secure facilities run by different governmental offices such as U.S. Naval Intelligence and the Department of Energy.

A human fingerprint is made up of ridges that take the shape of loops, arches, and whorls. Minutiae are the points on a fingerprint where a ridge ends or splits into two. The most promising minutiae points are extracted from an image to create a template, usually between 250 and 1,000 bytes in size [8, 15, 22-26, 133]. Fingerprinting is the most widely used biometric technology in the world, with a market share of 52% in 2003 [12]. Its small chip size, ease of acquisition and high accuracy have made it the most popular biometrics technology since the 1980s. However, some people may have fingerprints worn away due to hand work and some old people may have many small creases around their fingerprints, lowering the system's performance. In addition, the fingerprint acquisition process is sometimes associated with criminality, causing some users to feel uncomfortable with it.

A typical capacitive-based fingerprint sensor is shown in Figure 1-4 (a), where (b) exhibits a typical fingerprint image. Popular applications include the time attendance and employee management systems and physical access controls installed in the entrances of buildings.

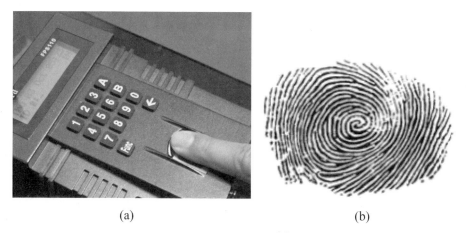

(a) (b)

Figure 1-4. (a) A fingerprint sensor, and (b) a typical fingerprint image.

Hand Geometry

There is only a small feature size in hand geometry, including the length, width, thickness and surface area of the hand or fingers of a user, as shown in Figure 1-5 (a) [6]. A project, called INSPASS (Immigration and Naturalization Service Passenger Accelerated Service System) [27], which allows frequent travelers to use 3D hand geometry, has been used in several international airports including Los Angeles, Washington, and New York. Qualified passengers enroll in the service to receive a magnetic stripe card with their hand features encoded. This allows them to simply swipe their card, place their hand on the interface panel, and proceed to the customs gate to avoid long airport queues. Several housing construction companies in Hong Kong have adopted Hand Geometry for the employee attendance record in their construction sites, as shown in Figure 1-5 (b). A smart card is used to store the hand shape information and employee details. Employees verify their identities by matching their hand features against the features stored in the smart card as they enter or exit the construction site. This measure supports control of access to sites and aids in wage calculations.

Hand geometry has several advantages over other biometrics, including small feature size, and low cost of computation as a result of using low resolution images [6, 28]. In spite of its current widespread use (10% market share in 2003 [12]), the current hand geometry system suffers from high cost and low accuracy [13]. In addition, uniqueness of the hand features is not guaranteed, making it unsuitable for use in one-to-many identification applications.

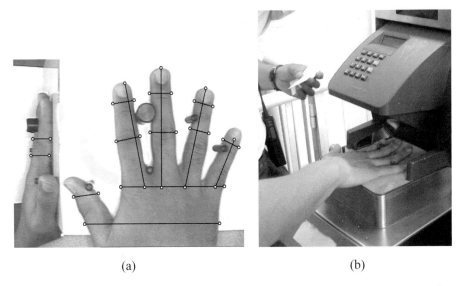

(a)   (b)

Figure 1-5. (a) The features of a hand geometry system (*Source: A. K. Jain*), (b) A hand geometry system operating in a construction site in Hong Kong.

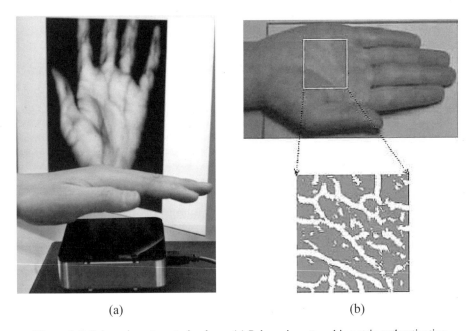

(a)   (b)

Figure 1-6. Palm vein pattern technology: (a) Palm vein pattern biometric authentication system, (b) sample vein patterns. (*Source: Fujitsu Laboratories Limited*)

Palm Vein Patterns

In August 2002 Fujitsu Laboratories Limited announced a new type of biometric authentication technology which verifies a person's identity by the pattern of the veins in his/her palms [29]. They claim that, except for their size, palm vein patterns are personally unique, and do not vary over the course of a person's lifetime. Infrared light is used to get the image of a palm when the hand is held over the sensor device. By testing 1,400 palm profiles collected from 700 people, the system achieved a false rejection rate of 1% and a false acceptance rate of 0.5% with an equal error rate of 0.8% [30]. The use of this technology would enable convenient biometric authentication for different types of applications such as log-in verification for access to sales, technical or personal data. Figure 1-6 (a) shows the Palm Vein Pattern Biometric Authentication System while (b) gives a sample of vein patterns. Since it is a new biometric technology, it does not yet have the public's confidence.

1.4   Why Palmprint?

Every biometric technology has its merits and limitations and no one technology is the best for every application domain. Nonetheless, as this Section outlines, palmprint does have advantages over other hand-based biometric technologies.

History of Palmprint

The history of the study of palmprints is ancient. From the start, palmprints have been used in fortune telling. Palmprints appeared in many Indian historical items including frescoes and vestiges. In China, about 3,000 years ago, palmprints for fortune telling were very popular. In the West Han Dynasty, there was a palmprint fortune telling book by Xu Fu called Sixteen Chapters of Physiognomy. It summarized some useful knowledge about palmprint, including types of palm lines, distribution of lines features, color of palm, and definitions of point features. China was the first country to use palmprints for forensic applications. In the Tang Dynasty, palmprints were applied to business contracts. In the Song Dynasty, palmprints were admissible evidence in courts of law. Palmprints were recognized as a means of measurement that uniquely represented a person.

Formation of Palmprint

To understand the formation of palmprints, we need to first understand its histology. Our skin is made up of three layers – epidermis, dermis, and subcutaneous tissue. The epidermis is the outermost layer of skin. It acts as a barrier to the external environment. The cells of the epidermis are called keratinocytes. These extend from the bottom layer of the epidermis to the top, building up a large amount of keratin and developing a tough outer shell. The epidermis of our palm may be as thick as 0.8 mm. That of most of our body is just 0.07 to 0.12 mm thick [131]. The thickening of the epidermis begins after birth, in response to the continuous pressure and friction.

Three types of lines can be found in a palm: flexure lines, tension lines, and papillary ridges [132]. The formation of these lines is related to the finger movements, the tissue structures, and the purpose of skins. For example, the flexure lines on our palm are the thickest and are effect skin hinges which open and close during grasping and gripping, following the movements of the underlying joints. These permanent creases are called principal lines throughout this book. In palmist terminology, these are called as life line, head line and heart line.

Principal lines

Although these are formed in the same way in everybody, there is a great deal of variability in these lines from person to person, as they grasp/grip objects in quite a different ways. The dimensions of the fingers/hands and the thickness of the palms/fingers of each person are also different, which makes the patterns of these lines even more variable. There are also a number of subsidiary flexure creases which vary in position in individual hands. From [132], it is believed that the life line is the first line developed in the embryo. It is the principal skin hinge for the thumb.

Wrinkles

Tension lines provide the basis for the wrinkling of the skin of a palm [132]. These occur all over the body to provide the skin with a certain amount of "stretchability" in directions corresponding with the natural demands of the region. As on the surface skin of an elbow or knee, the tension lines are arranged horizontally to allow for the movement of our arm and leg. However, when the skin loses its elasticity with advancing years, they form permanent wrinkles. Pronounced wrinkling is usually due to muscle activity. Some wrinkles are congenital while others, particularly on the palm, are acquired by a lifetime of muscular activity associated with certain grasp/grip activities of a hand.

Ridges

Papillary ridges are permanent thickenings of the epidermis, the outer cellular layer of the skin. They are raised above the general level of the skin, and are only to be found on the palmar surface of the hand and the sole of the foot. Their distribution corresponds with the principle areas of gripping and weight-bearing where they serve very much the same function as the treads on an automobile tyre. Ridges vary in coarseness on different parts of the hand. They are at their finest on the finger tips and at their coarsest on the rest of the fingers, with the palm of the hand being intermediate [132]. The papillary ridges are arranged in parallel formations, sometimes in curved series, sometimes straight. These ridges are visually like the texture of a piece of corduroy provide friction to assist the hands in grasping objects.

The second layer of skin is the dermis, which contains the structural elements of the skin, the connective tissue. The bottom layer of skin is the subcutaneous tissue containing fat cells. Discussions of the details of these two layers of skin are beyond the scope of this book.

Definition of a Palmprint

The palm is defined as the inner surface of our hand from the wrist to the root of fingers. A print is the impression made in or on a surface by pressure. A *palmprint* is defined as the skin patterns of a palm, composed of the physical characteristics of the skin patterns such as lines, points, and texture. Palmprints may be found on the surface of an object, mainly due to the perspiration. A dryer hand produces less observable "prints". The study of palmprints can be divided into two main streams, forensic and non-forensic. The "actual" palmprints found on a crime scene are useful for forensic investigation while non-forensic study may use imaging techniques to obtain observable human "palmprints" of a palm.

Palmprint authentication is a means of personal authentication that uses unique palmprint features, which may or may not be observable to the naked eye. It can be achieved by designing an appropriate algorithm capable of separating two persons by their palmprint features. Palmprints are rich in features: principal lines, winkles, ridges, singular points and minutiae points, as shown in Figure 1-7. Palmprints have a surface area much larger than a finger tip but are covered with the same kind of skin of a finger.

Features from Palmprints
Many features of a palmprint can be used to uniquely identify a person. Figures 1-7 illustrate an inked palmprint and an inkless palmprint from the same palm. Both of them were obtained using 500 dpi resolutions. The line features of the inked palmprint have a higher contrast while the inkless palmprint preserves the texture features. Apart from the effects of resolutions, six major types of features can be observed on a palm. The first four of them can be obtained from both inkless and inked palmprints, while the last two can be obtained only from inked palmprints at relative high resolution.

- *Geometry Features*: According to the palm's shape, we can easily get the corresponding geometry features, such as width, length and area.

- *Principal Line Features*: Both location and form of principal lines in a palmprint are very important physiological characteristics for identifying individuals because they vary little over time.

- *Wrinkle Features*: In a palmprint, there are many wrinkles which are different from the principal lines in that they are thinner and more irregular. These are classified as coarse wrinkles and fine wrinkles so that more features in detail can be acquired.

- *Datum Points:* Two end points called datum points are obtained by using the principal lines (see Figure 1-7). These intersect on both sides of a palm and provide a stable way to register palmprints. The size of a palm can be estimated by using the Euclidean distance between these end points.

- *Delta Point Features*: The delta point is defined as the center of a delta-like region in the palmprint. Usually, there are delta points located in the finger-root region. These provide stable and unique measurements for palmprint authentication.

- *Minutiae Features*: A palmprint is basically composed of the ridges, allowing the minutiae features to be used as another significant measurement.

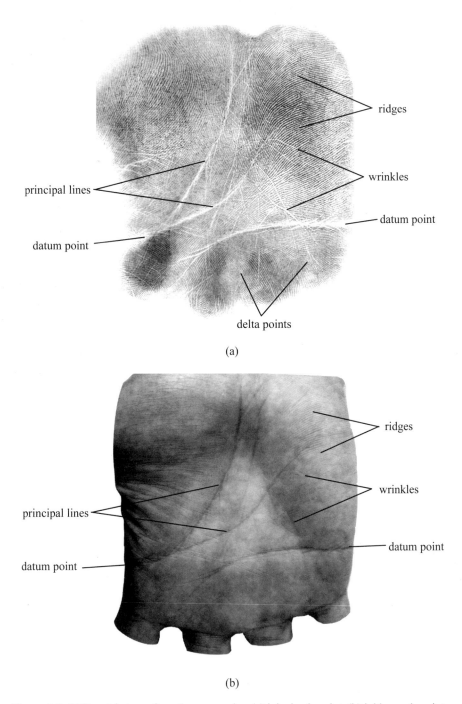

Figure 1-7. Different features from the same palm: (a) inked palmprint, (b) inkless palmprint.

Applications of Palmprints

Fortune Telling
Palmprints have traditionally been used to tell fortunes and to foretell the future. As the names of the principal lines implies, the life line is related to the life of a person, the heart line is related to the affections, and the head line is related to the intelligence of a person. The study of the principal lines is supposed to reveal our thoughts and feelings.

Health Diagnosis
Apart from fortune telling some research has focused on the health diagnosis according to changes in palm lines. Theories of Traditional Chinese Medicine and Holographic Medicine [129] mentioned that it is possible to diagnose diseases by observing palms. Many researchers have proved this point. Up to now, all research into palm diagnosis has been performed manually, which creates many difficulties. More research work should focus on developing an automated palm diagnosis system.

Personal Identification
The most promising application for palmprints is in security, as it is possible to obtain unique features from a palm for personal identification, just as with fingerprints. Palmprint recognition distinguishes between palms based on the lines, texture and points features in palmprint images. Features of different palms have a great variety. No two palmprints are completely identical.

Advantages of Palmprint
Apart from being feature-rich, palmprints have advantages over other hand-based biometric technologies:
(a) Compared with the fingerprint, the palm provides a larger surface area so that more features can be extracted.
(b) An individual is less likely to damage a palm than a fingerprint, and the line features of a palm are stable throughout one's lifetime.
(c) Small amounts of dirt or grease sometimes appear on an individual's finger, adversely affecting the performance of fingerprint verification. This problem does not arise in the extraction of palmprint features since we are using a comparative low resolution palmprint images (i.e. 100/125 dpi for offline and 75/150 dpi for online; details on palmprint images can be found in Chapter 2). Notice that the resolution of fingerprint images is 500 dpi.
(d) Compared with the 3D hand geometry technology, palmprints have more unique features that can be used for personal identification, so a better performance can be expected.
(e) Since palmprints use a much lower resolution imaging sensor compared with that of a fingerprints, with the result that computation is much faster in both the preprocessing and feature extraction stages.

In summary, many unique features of a palmprint image can be used for personal identification including principal lines, wrinkles, ridges, minutiae points, singular points and texture. All are useful in palmprint representation [31]. In view of these advantages, we have been working on palmprint research for personal identification in different real-time applications since 1996.

Perception on Palmprint Identifier

The applicability of a specific biometric technique very much depends on the application domain. Seven factors affect the determination of a biometric identifier in a particular application: universality, uniqueness, permanence, collectability, performance, acceptability, and circumvention [1]. These are defined as follows:

- *Universality* – each person should have the characteristic.

- *Uniqueness* – no two persons should be the same in terms of the characteristic.

- *Permanence* – the characteristic should be unchanging over a long period of time.

- *Collectability* – the characteristic should be quantitatively measurable.

- *Performance* – accuracy and speed of identification, and robustness of the system.

- *Acceptability* – refers to the extent to which people are willing to accept a particular biometric identifier in their life.

- *Circumvention* – refers to how easy it is to fool the system.

These factors can influence the selection of a biometric system. Table 1-1 summarises how biometrics experts perceive the common biometric technologies [16].

Table 1-1. Perception of five common biometric technologies by three biometrics experts (*Source: An Introduction to Biometric Recognition, IEEE* ©) [16].

Biometric identifier	Universality	Distinctiveness	Permanence	Collectability	Performance	Acceptability	Circumvention
DNA	H	H	H	L	H	L	L
Ear	M	M	H	M	M	H	M
Face	H	L	M	H	L	H	H
Facial thermogram	H	H	L	H	M	H	L
Fingerprint	M	H	H	M	H	M	M
Gait	M	L	L	H	L	H	M
Hand geometry	M	M	M	H	M	M	M
Hand vein	M	M	M	M	M	M	L
Iris	H	H	H	M	H	L	L
Keystroke	L	L	L	M	L	M	M
Odor	H	H	H	L	L	M	L
Palmprint	M	H	H	M	H	M	M
Retina	H	H	M	L	H	L	L
Signature	L	L	L	H	L	H	H
Voice	M	L	L	M	L	H	H

1.5   Arrangement of this Book

Palmprints have been accepted into the biometric family from the time it was first proposed by us in 1996 and this was reinforced with the publication of our research in 1998 (see [31]). Compared with other biometric technologies (Table 1-1), palmprint technologies can obtain a good results. In this book, we would like to summarize our palmprint work. Its fifteen chapters are in four parts, covering palmprint technology from the hardware components of the palmprint acquisition, to palmprint preprocessing algorithms, feature extraction, and matching.

PART I

Chapter One introduces recent developments in biometric technologies, some key concepts in biometrics, and the importance of developing a new biometric, the palmprint. Chapter Two focuses on the palmprint data: offline and online palmprints. Before the online palmprint acquisition device is developed, all the palmprints are obtained offline. Offline palmprints are obtained by placing an inked palm on paper, and then using a scanner to digitize the signal. Online palmprint acquisition is the most direct way to digitize a palmprint signal. A palmprint acquisition device developed by us was revealed. Different databases are built from these two methods for testing various experiments.

PART II

This part discusses preprocessing methods, feature extraction and matching, and the classification of offline palmprints. Chapter Three reports two registration methods for working on offline palmprints. The first method defines and extracts datum points on a palmprint which are useful for palmprint alignment. The second alignment method uses two invariant characteristics of a palmprint to handle the palmprints with rotation and translation. Chapter Four develops a line based feature extraction and matching strategy for offline palmprint verification and proposes a classification method based on the singular points found on the outside region of a palm. Chapter Five indicates a dynamic selection scheme for offline palmprint authentication which measures global texture features and then detects local interesting points. Palmprint patterns can be well described by textures: the texture energy of a palmprint possesses a large variance between different classes while remaining high compactness within the same class. The coarse-level classification using global texture features is effective and essential for reducing the number of samples for further processing at fine level.

PART III

This part has seven chapters, focusing on the palmprint acquisition requirement and design, preprocessing methods, feature extraction and matching, and the classification of online palmprints. Chapter Six presents three palmprint registration methods based on online palmprints. Two of these methods are designed to extract a fixed size square area as the central part palmprint sub-image for feature extraction, while the remaining one uses an inscribed circle as the central part of the palmprint sub-image. Chapter Seven reports a novel textured-based feature extraction method for personal authentication that uses low resolution online palmprint images. A palmprint is considered as a texture image, so an adjusted Gabor filter is employed to capture the texture information of palmprints. Combined with the effects of preprocessing techniques and rotational robustness of the filter, the matching process is translational and rotational invariant. Chapter Eight proposes a novel approach on palm line extraction and matching. A set of directional line detectors has been devised for effective palm line extraction. Palm lines are represented by their chain code to preserve the details of its structure. Then, palmprints are matched by matching the points on their palm lines. Chapter Nine discusses two novel algebraic features called Fisherpalms and Eigenpalms. Fisher's Linear Discriminant is used to project the palmprint image from the very high dimensional original palmprint space to the very low dimensional Fisher palmprint space, in which the ratio of the determinant of the between-class scatter to that of the within-class scatter is maximized. Similarly, eigenpalms method is developed by using the K-L transform algorithm. It can represent the principal components of the palmprints fairly well. The features are extracted by projecting palmprint images into an eigenpalms subspace. The weighted Euclidean distance classifier is applied. Finally, it can be seen that both fisherpalms and eigenpalms using the algebraic features from palmprint can achieve high recognition rates. Chapter Ten reports a novel feature extraction method by converting a palmprint image from a spatial domain to a frequency domain. A feature extraction method using Fourier Transform is first introduced. It can be seen that the similar palmprints resemble to each other when converted to frequency domain while different palmprints are separated far away from each other. Then, another transform-based approach using Wavelet signatures is reported. It is developed by characterizing a palmprint with a set of statistical signatures. The palmprint is transformed into the wavelet domain, and then cluster the predominant structures by context modeling according to the appearances of the principal lines in each subband. Chapter Eleven reveals a novel algorithm for palmprint classification using principal lines. A set of directional line detectors is devised for principal line extraction. Then, some rules for palmprint classification are presented. The palmprints are classified into six categories according to the number of the principal lines and their intersections. Chapter Twelve reveals a new idea on the online palmprint identification based on hierarchical multi-feature coding scheme. It can help to efficient and effective palmprint verification and identification in a large database by applying a coarse-to-fine matching strategy. There are four-level features defined: global geometry based key point distance, global texture energy, fuzzy "interest" line and local directional texture energy. In contrast to the existing methods using a fixed mechanism for feature extraction and similarity measurement, this approach (using multiple features) has higher performance.

PART IV

This part has three chapters. Chapter Thirteen introduces novel ideas for palmprint acquisition including a specially designed user interface and real-time palmprint acquisition capabilities, high quality palmprint image with reasonable price of the device are achieved. The fastest time to obtain a palmprint image at spatial resolution of 150 dpi is $0.09s$. Our proposed system is able to obtain features from a palm including principal lines, wrinkles and ridge texture. In Chapter Fourteen, we first present a novel biometric system based on the palmprint. It can accurately identify a

person in real time and is suitable for various civil applications such as access control. Experimental results including accuracy, speed and robustness demonstrate that the proposed palmprint authentication system is comparable with other hand-based biometrics systems, such as hand geometry and fingerprint verification system and is practical for real-world applications. Chapter Fifteen provides a brief summary of each chapter, then, discusses future palmprint research. It explores different research aspects including higher performance (in terms of accuracy rate and larger number of users), distinctiveness of palmprints, permanence of palmprints, palmprint databases, application-related issues, and privacy concerns.

2 PALMPRINT DATA

Data acquisition is the first stage of every biometric system, and palmprints are no exception. In this chapter, we first discuss two palmprint data acquisition methods: offline and online. Before the online palmprint acquisition device is developed, all the palmprints are obtained offline. Offline palmprints are obtained by pressing an inked palm onto paper, and then using a scanner to digitize the signal. Online palmprint acquisition is the most direct way to digitize palmprint data. Section 2.2 reports different palmprint acquisition devices developed by us. Section 2.3 introduces the offline palmprint databases. Section 2.4 reports two online palmprint databases. Section 2.5 provides a summary of this chapter.

2.1 Palmprint Data Acquisition

In general, a biometric system consists of the following four stages: 1) Data acquisition 2) Signal/Image preprocessing 3) Feature extraction and 4) Feature matching, as described in Chapter One. This section will focus on palmprint data acquisition. There are two acquisition methods: offline and online. Offline palmprint acquisition means that the palmprint data is not directly input to the computer. Instead, it is first inked on a sheet of paper, and then a scanner is used to digitize the data and it is stored in a computer for further processing. Online palmprint acquisition is the most direct way to digitize palmprint data, obviating the need for a third medium like paper. It can be accomplished by using a scanner to scan a palmprint directly, or using a video camera to obtain the palmprint data.

Offline Method
We started our palmprint research in 1996 using inked images, as shown in Figure 2-1. The offline method involves collecting samples by inking a user's palm and pressing it onto a sheet of white paper. After the ink has dried, the palmprint image on the paper is digitized with a scanner and stored in the personal computer [18]. Clearly, this method is not, however, suitable for real time application such as physical access control. Apart from the number of steps involved, the quality of the palmprint image is not satisfactory because palmprints may be affected by the amount of ink used. Both too much ink and too little ink produce unsatisfactory palmprints.

Figure 2-1. An inked palmprint image from offline acquisition method.

Online Method

We designed the world's first online palmprint capture device in December 1999, at The Hong Kong Polytechnic University [10]. This was the first attempt to make an online device for palmprint research. The device was made using a plastic box, a light source, a mirror, a glass plate and a CCD camera, as shown in Figure 2-2. Repeated testing showed that the image formed through the mirror was not as good as a direct reflection because the second surfaced mirror creates a ghost image. The glass plate used to hold the palm distorted the surface of the skin of the palm so that palm lines were not good enough for feature extraction. A better device was needed in order to get a better image quality for further processing.

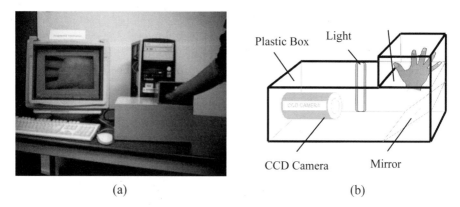

(a)   (b)

Figure 2-2. The first online palmprint acquisition device: (a) appearance, and (b) architecture.

To develop a real-time, online palmprint identification system requires a palmprint scanner that can quickly capture high quality palmprint images. This is fundamental to palmprint research, yet limited research efforts have been put into designing and implementing a palmprint acquisition device. Such a device must be able to online acquire a good quality palmprint image. Users should also find the interface comfortable to use. Let us look at the specific criteria for our final version of the palmprint acquisition device, its user interface requirements and optical system requirements.

I. User Interface Requirements
● *Easy and intuitive to use* – Users should feel comfortable when they are using the system since the level of user acceptance is very important in customer-based applications.
● *Size* – The size is dependent on applications, but it should be as small as possible.

II. Optical System Requirements
● *Image quality* – High quality images must be obtained such that it is possible to perform the subsequent image processing.
● *Real time processing* – The response time of the system should be fast enough to provide a development platform for practical applications.

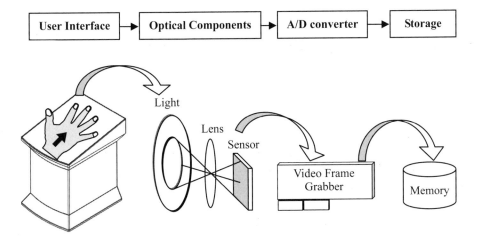

Figure 2-3. The structure of the palmprint acquisition system.

The structure of the proposed system is shown in Figure 2-3. There is a user interface for the input of a palm. A set of optical components work together to obtain the data from the palm. The analog signal is converted into a digital signal using an A/D converter and the digitized signals are stored in the system's main memory.

The user interface has a specially-designed flat platen surface for holding the palm in alignment. It acts as the input channel for the user and system, and for the acquisition of the palmprint data from the user interface. The optical components (light source, lens, and CCD sensor board) and the A/D converter (video frame grabber) are the heart of the palmprint acquisition device. The light from the object (palm) passing through the aperture is converged by the lens to form an image. A ring light source provides a high intensity white light source. This can increase the contrast of the palmprint features when palm skin surfaces are uneven. The optical system is fabricated in a controlled environment in order to minimize the effects of ambient light. The image quality can be affected by any of these components. A CCD sensor transforms the photon signal into an electrical signal. A video frame grabber processes the analog signal from the CCD sensor and an Analog to Digital Conversion (ADC) takes place in the on-board chip. The palmprint image is then stored in the system memory. In addition, an interface board was designed to allow the user to communicate with the system via a keypad and display unit. The proposed system uses an Intel Celeron 950 MHz CPU, 128 MB RAM and storage of 32MB Disk on Module. Detailed implementation of the palmprint acquisition system including the requirement analysis, system design, user interface design, and system evaluations are discussed in Chapter 13. We have built a prototype of the system, with the recognition engine design [65] to provide access control to our laboratory; details are provided in Chapter 14.

2.2 Different Online Palmprint Acquisition Devices

Four versions of the online palmprint acquisition have been designed so far: 1) an L-shaped design, 2) a long-tube horizontal design, 3) a long-tube vertical design, and 4) an enhanced short-tube design (the final design). Different design stages have focused on different objectives such as image quality, user acceptance, and so on. Each version has been a significant improvement. Different versions look very different but internal configurations have also changed considerably in order to improve palmprint image quality and to support higher user acceptance. These changes include the size and shape of the device, the optical path, the lighting used, and the user interface design.

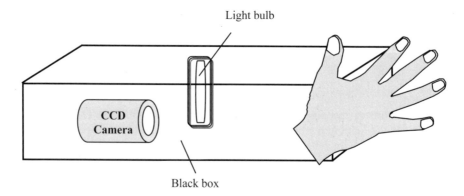

Figure 2-4. Second version palmprint acquisition device: long-tube horizontal device.

First Design –L-Shaped

The first generation of our design is an L-shaped device with a internal mirror for reflecting light from the palm to the CCD sensor [10], (see Figure 2-2). The mirror cuts the optical axis horizontally and vertically. This reduced the height of the device but the mirror could cause degradation of the image quality.

Second Design – Long-Tube Horizontal

A long-tube horizontal device (Figure 2-4) is designed to address problems arising from the previous L-shaped design. Since the most important issue in a palmprint acquisition device is image quality, the optical system uses the traditional straight-through optical axis. We also remove the glass plate so it would be possible to obtain palmprint features more directly without distortion by the glass and mirror. For the lighting design, we use a light bulb in the optical system to illuminate the palm. This approach greatly improves image quality.

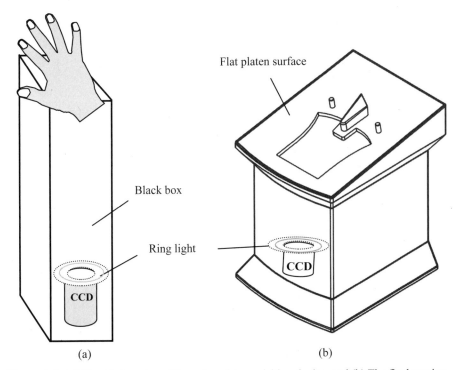

| (a) | (b) |

Figure 2-5. (a) The third version of the palmprint acquisition device, and (b) The final version of the palmprint acquisition device.

Third Design – Long-Tube Vertical

Users found that the horizontal arrangement of the device in the second design was inconvenient, so we made it upright, ensuring uniform illumination of the palm by changing the light source from a light bulb to a fluorescent ring light (Figure 2-5 (a)). These changes support users in a more intuitive use of the device, and produce a more uniformly illuminated palmprint image.

For palmprint acquisition, we designed a user interface called a flat platen surface. It is in appropriate size, comfortable to use and can accommodate most people. In order to help users correctly place their palm, we added pegs as shown in Figure 2-6 (a). The dotted line area is cropped from the flat platen surface to allow the CCD camera underneath to acquire the palmprint image. We carried out tests to determine a suitable size for the flat platen surface and positioning of the pegs. With the current design, we can obtain images from large or small hands, as shown in Figures 2-7. As the images shown, all the important line features are obtained.

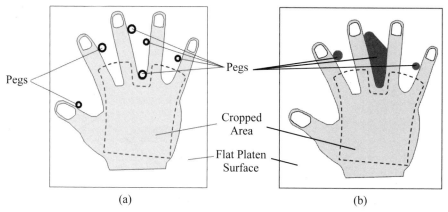

(a)   (b)

Figure 2-6. Flat platen surface design: (a) first attempt, and (b) the enhanced designed.

Final Design – Enhanced Short-Tube

Previous designs proved the lighting environment of the device to be effective so we were able to use a 6 mm focal length lens to reduce the height of the optical path from 240 mm to 130 mm. Further reduction in focal length could distort the captured image.

For the user interface design, our experience has shown that a proliferation of pegs on the flat platen surface can confuse users as to how they should place their palms. The ultimate design takes account of this, and the use of pegs balances user convenience and system stability. As shown in Figure 2-6 (b), the flat platen surface requires only three pegs: a large triangular peg to guide the middle and ring fingers; one round-shaped peg on the left to guide the index finger; and one peg on right to guide the little finger. We provide a clear and easy-to-use user interface and a stable palmprint alignment positioning mechanism. The design of the flat platen surface is thus satisfactory from both the user and system perspectives.

The final version of the device was formed by applying this enhanced flat platen surface along with the best light source (ring fluorescent light), as shown in Figure 2-5 (b). In our proposed palmprint acquisition device, we use a 1/2" lens format and a 1/3" CCD sensor format so that lens distortion near its rim is eliminated. The experimental results illustrate that the palmprint images captured using our final acquisition device satisfy the requirements.

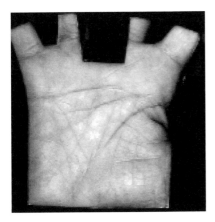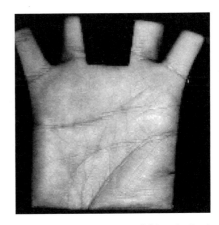

Figure 2-7. Palmprint images in a small hand and a large hand by our acquisition device in Figure 2-5 (a).

2.3 Offline Palmprint Databases

When we start the palmprint research, we should collect palmprint images to form an offline palmprint database so that different algorithm tests can be performed. Offline palmprints are obtained by thoroughly inked the inner surface of a palm and then put a sheet of white paper on the palm, ensuring the paper makes full contact with the inner surface, and then scan the paper to get a digital palmprint image. In part two of this book (Chapters 3-5), we use three offline palmprint databases. They were collected from three different stages, with different resolutions and image dimension settings, and the details are summarized in Table 2-1.

STAGE ONE
In 1996 we collected palmprints of 20 individuals – 10 prints from both left and right palms to form a database of a total of 400 palmprint images, called **PolyU-OFFLINE-Palmprint-I**. Ridge patterns and singular points can be obtained from high resolution images, i.e. 500 dpi, but the processing time and power for this resolution is huge, i.e. a dimension of 2,000 × 2,000 with around 4 MB in size for 500 dpi resolution palmprints. We used 100 dpi at 400 × 400 to digitize these inked palmprints paper sheets in order to reduce the computation burden. The database was designed especially to test *datum point determination* (see Chapter 3 Section 3.2), and *line matching* (see Chapter 4) algorithms. Low resolution palmprint are adequate for use with these algorithms. Figure 2-8 shows two typical palmprint images from this database.

STAGE TWO
Having tested different algorithms on the **PolyU-OFFLINE-Palmprint-I** database, we collected another set of offline palmprints which had different image sizes and dimensions. Again, as high resolution palmprints need higher computational power and increases the complexity, we chose to use low resolution, i.e. 125 dpi at 432 ×

432, to digitize these inked palmprints paper sheets. This stage of palmprint collection started in 1999. We collected palmprints of 100 individuals with two prints from the right palm. We formed a new database totally 200 palmprint images, called **PolyU-OFFLINE-Palmprint-II**. Some new algorithms were then tested using this new set of offline palmprints, such as the method *inked image alignment* (see Chapter 3 Section 3.3) and the *layered palmprint identification* (see Chapter 5). Figure 2-9 shows two typical palmprint images from this database.

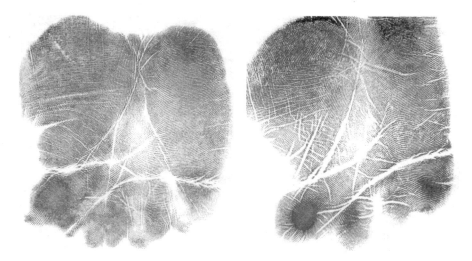

Figure 2-8. Two typical palmprint images from **PolyU-OFFLINE-Palmprint-I**.

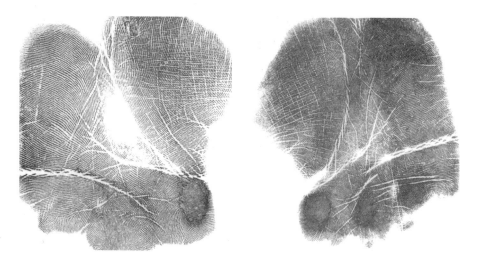

Figure 2-9. Two typical palmprint images from **PolyU-Offline-Palmprint-II.**

STAGE THREE

Although we have built up two offline palmprint databases, we proposed another new idea for offline palmprint classification, so that we needed to build a third database with more palmprints inside. Palmprint classification requires fine detail, i.e. singular point, on the outer region of a palm. So, we obtain this new database by inking only the outer region of a palm on a sheet of paper, then use 500 dpi resolutions to digitize these inked palmprints paper sheets. In this stage, we collected 354 palmprints, with one sample per person. This database is called **PolyU-OFFLINE-Palmprint-III**. It was applied to the method called *singular point for classification* (see Chapter 4). Figure 2-10 shows two typical palmprint images from this database.

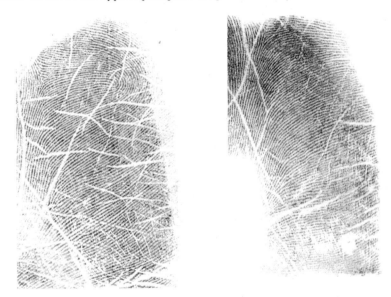

Figure 2-10. Two typical palmprint images from **PolyU-Offline-Palmprint-III**.

Table 2-1. Attributes of the three offline palmprints databases.

Database	Stage 1 **PolyU-OFFLINE-Palmprint-I**	Stage 2 **PolyU-OFFLINE-Palmprint-II**	Stage 3 **PolyU-OFFLINE-Palmprint-III**
No. of persons	20	100	354
Samples per person	20	2	1
Dimension	400×400	432×432	$1,250 \times 2,000$
Resolution (dpi)	100	125	500
Graylevel	256	256	256
Total no. of images	400	200	354

2.4   Online Palmprint Databases

As mentioned in Section 2.1, different palmprint acquisition devices have been designed in our palmprint research. They can be used to collect different palmprint images. In this book, Parts three and four deal with two online databases captured using these acquisition devices.

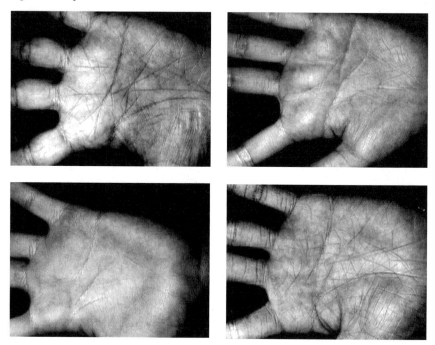

Figure 2-11. Some typical palmprint images from **PolyU-ONLINE-Palmprint-I**.

STAGE ONE

When we had successfully designed the first generation online palmprint acquisition device (an L-shaped design, Section 2.1), we started to collect palmprints from different persons for the experimental testing. After some tests, we found that the image quality was not satisfactory. This motivated us to design a better palmprint acquisition device. The second version of palmprint acquisition device (see long-tube horizontal design in Section 2.1) was built. We then prepared to collect our first online palmprint database, **PolyU-ONLINE-Palmprint-I**, for further testing. This database has different individuals from the students and staff at The Hong Kong Polytechnic University. This device can capture only right palms. The image size is 320×240 at 75 dpi. We collected about 500 persons, each providing 6 to 10 right palms. Figure 2-11 shows some typical palmprint images acquired using the second generation online palmprint acquisition device. Since the collection of this database was time consuming, the number of palmprint images from each individual may vary between six and ten. Palmprint database collection continued until June 2001.

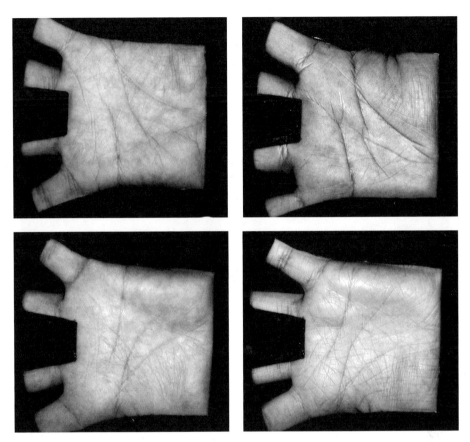

Figure 2-12. Some typical palmprint images from **PolyU-ONLINE-Palmprint-II**.

STAGE TWO

To obtain higher quality and more stably oriented palmprints, we developed the third generation of online palmprint acquisition device (a long-tube vertical design, see Section 2.1). This has a user interface called flat platen surface for palmprint acquisition. On the flat platen surface, there are pegs which guide the correct placement of a palm during the acquisition process. The lighting design was also improved so that the images acquired are better than the previous one. The palmprint image formed is 768 × 568 at 150 dpi. The image resolution can be further reduced to 75 dpi (1/2 of the original) by adjusting the system parameter, i.e. the image size is 384 × 284 pixels at 75 dpi. Since then, we started to collect our second online palmprint database, called **PolyU-ONLINE-Palmprint-II**.

From the very beginning, we obtained palmprints in 384 × 284 at a resolution of 75 dpi. As our objective was to use low resolution palmprint images for the algorithm design we thought there was no need to use so high a resolution for the palmprint personal identification. However, we found that the larger size (i.e. 768 × 568 at 150 dpi) palmprints may be useful for some special applications such as palm line extraction for palmistry, so we changed the system parameter so that all the palmprints acquired were in 768 × 568 at 150 dpi. Thus we provide two different online palmprint resolutions for researchers to use. Figure 2-12 shows some typical palmprint images acquired using the third generation online palmprint acquisition device (**PolyU-ONLINE-Palmprint-II**).

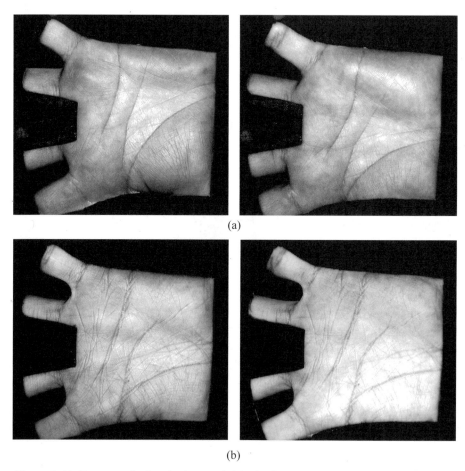

(a)

(b)

Figure 2-13. Two sets of palmprint images, left is the first time while right is the second time acquisition.

In this collection, 200 individuals provided their palmprints. The subjects mainly consisted of volunteers from the students and staff at The Hong Kong Polytechnic University. In this dataset, 131 people are male, and the age distribution of the subjects is: about 86% younger than 30 years old, about 3% older than 50, and about 11% aged between 30 and 50. We collected the palmprint images on two separate occasions, at an interval of around two months. On each occasion, the subject was asked to provide about 10 images each of the left and right palms. We instructed the subjects in how to use our capture device. Each person provided around 40 images so that our database contained a total of 8,025 images from 400 different palms. In this database, 580 palmprints out of 8,025 palmprints are captured in 384 × 284 at 75 dpi. The remainder are 768 × 568 at 150 dpi. Figure 2-13 shows two sets of palmprint images that were captured at different occasions.

We changed the light source and adjusted the focus of the CCD camera so that the images collected on the first and second occasions could be regarded as being captured using two different palmprint devices. The average interval between the first and second occasions was 69 days. The maximum and minimum time intervals were 162 days and 1 day, respectively. Table 2-2 summarizes the detailed configurations of both **PolyU-ONLINE-Palmprint-I** and **PolyU-ONLINE-Palmprint-II**, which are used in Chapters 6 to 14.

Currently, we are collecting the third online palmprint database using the newest (i.e. the 4th generation, final version) palmprint acquisition device. Since the number of users obtained at this moment is very small, we do not publish details of this database in this book.

Table 2-2. Attributes of the two online palmprints databases.

Database	Stage 1 **PolyU-ONLINE-Palmprint-I**	Stage 2 **PolyU-ONLINE-Palmprint-II**		
No. of person	500	200		
Samples per person	6 to 10 (Right)	Left	1st time	10*
			2nd time	10*
		Right	1st time	10*
			2nd time	10*
Dimension	320 × 240	384 × 284 / 768 × 568 *		
Resolution (dpi)	75	75 / 150 *		
Gray levels	256	256		
Total no. of images	4,523	8,025		

* Among 8,025 palmprints, 580 palmprints are captured in 384 × 284 at 75 dpi; some persons provide more than 10 samples while some provide less, but most of them have 10 samples for each hand at each occasions.

2.5   Summary

In this chapter, we discussed the methods of palmprint acquisition, offline and online approaches. Before the online palmprint acquisition device is developed, all the palmprints are obtained offline. Offline palmprint acquisition means that the palmprint data is not directly input to the computer. Rather, it is first inked on paper, and then a scanner is used to digitize the signal and store it in a computer for further processing. Online palmprint acquisition is the most direct way to digitize palmprint data, without the need for a third medium like paper. Four online palmprint acquisition devices developed by us are revealed. Different databases were built, in which palmprint images were obtained using both methods, for the experimental testing. We also provided information on different palmprint databases, including the image size, number of images, and resolution. In later chapters, different algorithms are tested on the palmprint databases.

PART II

OFFLINE METHODOLOGIES

3 PALMPRINT PREPROCESSING

Palmprints may exhibit some degree of distortion as palmprint signals may be captured at different times, and under varying conditions of temperature, humidity, and brightness, etc. Apart from noise reduction and the smoothing of palmprint images, palmprint preprocessing involves correcting distortions and putting all palmprints under the same coordinate system so that the expected area of each palmprint can be extracted for use in feature extraction and matching. This chapter discusses the palmprint preprocessing technique using offline palmprint images. Section 3.1 provides some notations and definitions. Section 3.2 reports a method using the datum point for the palmprint alignment. Section 3.3 demonstrates a coordinate system for palmprint alignment on inked palmprints. Section 3.4 summarizes this chapter.

3.1 Introduction

Notations and Definitions

We have defined a palmprint in Chapter 2, but there are still some more specific definitions and notations in detail for the palmprint images. In general, we can define three principal lines on a palm, called as the lines of the life, heart and head (see Lines 1, 2, and 3 in Figure 3-1). They are unique, unchanging, and cannot be forged. In addition, two end points, **a** and **b**, can be obtained by the principal lines, which intersect both sides of a palm. Since the stability of the principal lines, the end points and their midpoint **o** remain unchanged day in and day out. Some significant properties can be proposed by the above findings:

a. The locations of the end points and their midpoint are rotation invariant in a palmprint,

b. A two-dimensional right angle coordinate system can be established uniquely, of which the origin is the midpoint **o** and the x-axis passes through the end points,

c. The size of a palm can also be uniquely estimated by both the Euclidean distance between two end points and the length of their perpendicular bisector in the palm (see **c-d** in Figure 3-2),

d. A palm can be divided into three regions, named as finger-root region (I), inside region (II) and outside region (III), by a connection between the end points and its perpendicular bisector (see Figure 3-1), and

e. There are some Delta Point Features (see Points 1-5 in Figure 3-2) which is defined as the center of a delta-like region in the palmprint. It is shown that there always are some delta points located in finger-root region and outside region. This palmprint is digitized using 500 dpi resolution, so that it is possible to obtain these features and well establish the stability and uniqueness.

As a result, the end points and their midpoint can be taken as an important registration in the palmprint automatic matching because of their invariable locations.

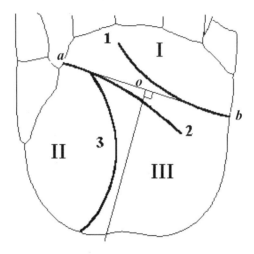

Figure 3-1. Definitions of a palmprint: principal lines (1–heart line, 2–head line and 3–life line), regions (I–finger-root region, II–inside region and III–outside region) and datum points (*a*, *b*–endpoint, *o*–their midpoint).

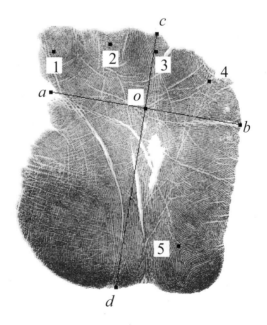

Figure 3-2. Geometry features and delta point features of a palmprint, where *c-d* is the perpendicular bisector of line segment *a-b* and points *1-5* are delta points

3.2   Datum Point Registration

The goal of datum point determination is to achieve the registrations in palmprint feature representation and matching. As a prime but important process in the palmprint authentication, it is demanded as simple and effective as possible. The basic idea of datum point determination is to locate the endpoint of each principal line. According to the given regularity, the principal lines and their endpoints are accurately detected by using the directional projection algorithm.

Directional Projection Algorithm

It is widely known that projection is a simple yet effective method of line segment detection along some orientation [34].

Let $\mathbf{F}$ be an $M \times N$ gray scale image and $f(i, j)$ be the gray level of pixel (i, j) ($i = 0$, 1, ..., M-1; $j = 0$, 1, ..., N-1). Without loss of generality, we consider that the projective angle α is measured clockwise from the i-axis and pixel (i', j') is a pixel in $\mathbf{F}$. Then, an x-y right angle coordinate system can be established, of which pixel (i', j') is the origin and the orientation of x-axis is that of the projection (see in Figure 3-3). In the x-y coordinate system, a $(2m+1) \times (2n+1)$ pixels in size subimage, $\mathbf{F}_1$, can be obtained, and $f(x, y)$ is the gray level of pixel (x, y) ($x = -m, -m+1, ..., 0, ..., m$-1, m; $y = -n, -n+1, ..., 0, ..., n$-1, n). As a result, the correspondence between this pair of coordinate systems is denoted as follows:

$$i = i' + \cos(\alpha + \beta)\sqrt{x^2 + y^2}$$
$$j = j' + \sin(\alpha + \beta)\sqrt{x^2 + y^2} \ , \tag{3-1}$$

where $\beta = \tan^{-1}(y / x)$.

In $\mathbf{F}_1$, obviously, the directional projection of this subimage is

$$p(y) = \sum_{i=-m}^{m} f(x, y) . \tag{3-2}$$

The smoothed set $q(y)$ is calculated as

$$q(y) = \frac{1}{2w+1} \sum_{k=-w}^{w} p(y + k) . \tag{3-3}$$

Then, pixel $(0, y_0)$, where

$$y_0 = \{k \mid q(k) = \max_{y} q(y)\}, \tag{3-4}$$

is detected and the corresponding pixel in $\mathbf{F}$ is defined as pixel (i_0, j_0), where

$$i_0 = i' + y_0 \cos(\alpha + \frac{\pi}{2})$$
$$j_0 = j' + y_0 \sin(\alpha + \frac{\pi}{2}) \ . \tag{3-5}$$

Here, this basic algorithm is classified as four forms by the different projection orientations: horizontal projection ($\alpha = 0°$), projection with 45 degrees ($\alpha = 45°$), vertical projection ($\alpha = 90°$) and projection with 135 degrees ($\alpha = 135°$).

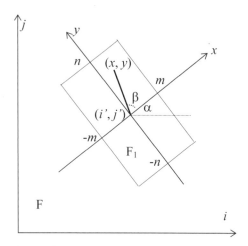

Figure 3-3. The correspondence of two coordinate systems in directional projection algorithm.

Properties for Principal Lines
From the study of the principal lines, their properties can be obtained as follows:
(1) Each principal line meets one side of palm at approximate right angle when it flows out the palm;
(2) The life line is located at the inside part of palm, which gradually inclines to the inside of palm in parallel at the beginning;
(3) In most cases, the life line and head line flow out the palm at the same point;
(4) The endpoints (**a**, **b**) are closer to fingers than wrist.

Based on the projective correspondence (see Figure 3-4), it is clear that the pixel (x, y) calculated by the basic algorithm belongs to the principal line if the orientation of directional projection follows the principal line and the parameter w in Equation (3-3) equals the half width of a principal line. However, the pixel on the other different kinds of the lines cannot be determined by means of the conditions given above. This is because the projection of line which is not in parallel with the projective orientation or is shorter than the principal line would be less than that of principal line. In addition, a thin line might map to the maximum value in the directional projection, $p(y)$, but it could be reduced after smoothing. So, the basic algorithm can be used to locate the pixel only on the principal line while the directional projection along the principal line is maximized in a palmprint subimage.

Datum Point Determination
To apply the above rules to get the endpoints, two processing stages: i) principal line detection stage, and ii) principal line tracing stage, should be defined. Without loss of generality, the images of person's right palm are employed and all palmprints are considered as fingers upturned so as to describe the method of datum point determination simply. The track of the datum point determination is shown in Figure 3-5.

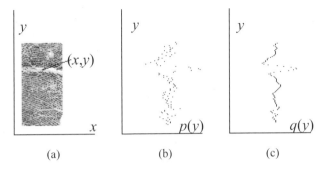

Figure 3-4. Computation of the directional projection: (a) original image; (b) directional projection along principal line;, and (c) directional projection after median filter.

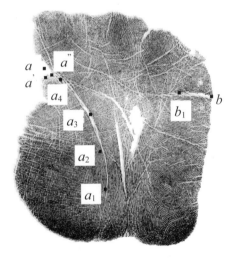

Figure 3-5. Track of datum point determination by using the directional projection algorithm.

i) Principal Line Detection

The purpose of this stage is to obtain the points, that belong to the heart line or the life line. The heart line detection is easier than that of life line because there is only one principal line in the outside part of palm. After the edges of outside and topside are detected, a pixel which has the suitable offsets to two edges is adopted in palmprint. Point b_1, which belongs to heart line, is located by the horizontal projection algorithm. However, the detection of life line cannot use the horizontal projection in the inside part of palm because the head line is determined sometimes instead of the life line while their flowing out palm at the different points. Therefore, another subimage close to wrist is processed by the vertical projection algorithm and point a_1 on the life line is calculated.

ii) Principal Line Tracing

The main idea is to estimate a pair of endpoints by tracing principal lines. According to the peculiarities of heart line, the horizontal projection algorithm is easy to locate endpoint b to a subimage where there is a pixel situated at the outside edge with the same level as pixel b_1. However, the detection of endpoint a is different from that of endpoint b because the life line is a curve from pixel a_1 to point a. As a result, the details of point a determination are shown below:

Step 1: Point a' is defined by the straight line, which is a connection between endpoint b and b_1, intersecting the inside edge of palm.

Step 2: Pixels a_2 and a_3, which trisect the vertical distance between pixel a_1 and point a', are obtained in proper order along the life line by the vertical projection algorithm.

Step 3: Pixel a_4 belonging to the life line is denoted. The horizontal distance from pixel a_3 to the inside edge of palm is calculated and a new point can be located, from which both horizontal and vertical distances are half of above distance to pixel a_3. A subimage around this new point is processed by the directional projection algorithm with 135 degrees.

Step 4: Point a'' is detected by the straight line, which flows out from pixel a_4, intersecting the inside edge of palm at 22.5 degrees.

Step 5: Endpoint a is estimated in the subimage with point a'' by the horizontal projection algorithm.

Based on the endpoint a and b, it is simple to determine the midpoint o. For the left palmprint's datum point detection, the similar steps are used except that the directional projection algorithm in Step 3 is replaced by that with 45 degrees. Obviously, the proposed method is also suitable for the palmprint under rotation while the outside edge of palm is detected and rotated to approximate vertical.

Experimental Result

The method of datum point determination has been tested on the offline palmprints database, **PolyU-OFFLINE-Palmprint-I**. We picked 300 palmprint images from the database randomly, where 286 are found to be in excellent agreement with the manual estimate. In particular, this algorithm is used to estimate the datum points with 60 special palmprint images, which 20 belong to under rotation, 35 to incomplete and 5 to those life line and head line flow out the palm at different points. Some typical palmprint images are shown in Figure 3-6 and the results of the datum points detection from special palmprint images are given in Table 3-1.

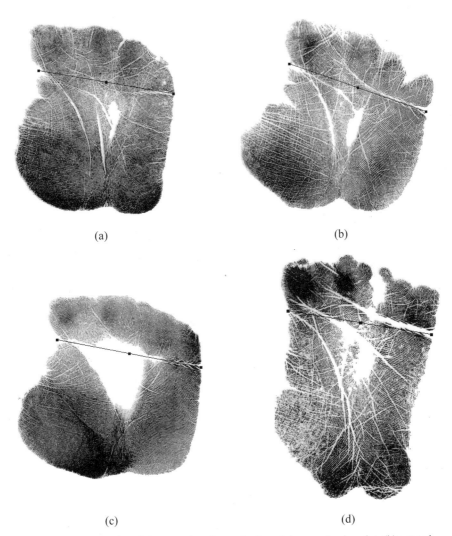

(a) (b)

(c) (d)

Figure 3-6. Examples of datum points determination: (a) normal palmprint; (b) rotated palmprint; (c) incomplete palmprint; and (d) the life line and head line unintersection.

Table 3-1. Experimental results of datum point determination in special palmprint images

Classification	R	I	U
Experiment images	20	35	5
Accurate determination images	19	33	4
The rate of accuracy (%)	95	94	80

Note: R = rotated image; I = incomplete image;
U = life line and head line unintersetion.

3.3   Inked Image Alignment

The objective of palmprint alignment is to put all the palmprints into the same location in their images. We proposed a method to align inked palmprint image, which has four main steps [137]: A) defining a coordinate system, B) determining the Y-axis, C) determining the origin, and D) rotating and translating the original image.

A) Coordinate System Definition

In fact, the outside boundary of a palmprint is usually clear and stable which can be described by a straight line. This means that the outside boundary can be defined as the Y-axis, and the intersect point between the outer boundary and principle line given in [18] as the origin. Therefore, such a two-dimension right angle coordinate system can be setup, as shown in Figure 3-7. Using this coordinate system, we may move all the palmprints to a certain position with the same direction in their images so that the comparisons of palmprint are rotation and translation invariant.

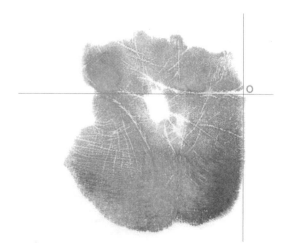

Figure 3-7. A two-dimension right angle palmprint coordinate system using two invariant features - outer boundary direction (Y-axis) and end point of principle line (origin).

B) Y-axis Determination

The Y-axis in the original image is denoted as:

$$y = ax + b,$$ (3-6)

where

$$a = \bar{y} - b\bar{x} \quad \text{and} \quad b = \frac{l_{xy}}{l_{xx}}.$$ (3-7)

Each variable is defined as follows:

$$\bar{x} = \frac{1}{n} \sum_{i=1}^{n} x_i, \quad \bar{y} = \frac{1}{n} \sum_{i=1}^{n} y_i$$ (3-8)

$$l_{xx} = \sum_{i=1}^{n}(x_i - \bar{x})^2, \quad l_{xy} = \sum_{i=1}^{n}(x_i - \bar{x})(y_i - \bar{y}) \qquad (3\text{-}9)$$

where $(x_i, y_i)(i=1,\dots,n)$ are points on the edge of the palmprint's outer boundary. Figure 3-8 illustrates the Y-axis determined by the given method. Figure 3-8 (a) is the original image and (b) is the binarized palmprint image on which boundary tracing is performed. Figure 3-8 (c) shows the outer boundary and a straight line, which describes its direction. Such a straight line thereafter is defined as the palmprint's Y-axis.

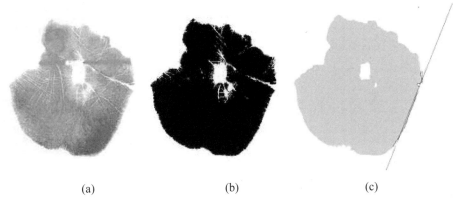

(a) (b) (c)

Figure 3-8. Main process for Y-axis determination

C) Origin Determination
Origin determination involves the detection of the end points of principle lines on a palm, which is based on the original grayscale image. After defined the direction of a palmprint, a projection is conducted to the Y-axis to obtain the position of an end point which has the largest energy. Such a process of finding origin is shown in Figure 3-9. Note that Figure 3-9 (a) is the rotated palmprint image and the rectangle on upper right corner is used to extract a sub-image on which the origin of the palmprint coordinate system is determined. The sub image extracted from Figure 3-9 (a) is enlarged in Figure 3-9 (b). As a horizontal projection map of Figure 3-9 (b), we can get the largest energy on the intersect point between the heart line and the outer boundary in Figure 3-9 (c).

D) Image Rotation and Translation
The image rotation and translation formula are:

$$x' = x\cos\theta + y\sin\theta + d_1, \qquad (3\text{-}10)$$

$$y' = -x\sin\theta + y\cos\theta + d_2. \qquad (3\text{-}11)$$

Having determined the Y-axis and origin, we can rotate the palmprint image so that all inked palmprint are aligned in the correct position.

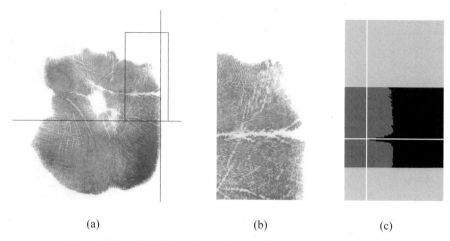

(a)   (b)   (c)

Figure 3-9. Illustration of palmprint origin determination.

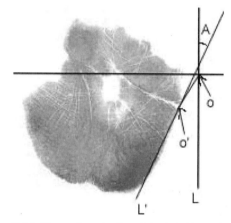

Figure 3-10. Illustration of palmprint origin determination.

3.4 Experimental Result

The palmprint images for the testing are based on all of the images from the offline palmprints database, **PolyU-OFFLINE-Palmprint-II**, i.e. 200 images from 100 different individuals. Experiments have been carried out to verify the performance of the proposed alignment method. The testing is conducted step by step as follows:

i) Before the alignment, the origins and directions of all palmprints are manually determined. In Figure 3-10, L' is a straight line on a fixed location; o' is a fixed point on L'; L is the palmprint's Y-axis and o is the origin determined manually. Also, A is defined as the angle between L and L' and D as the distance between o and o'. We calculate the As and Ds in all palmprints to obtain their average A (which we call 'direction deviation') and average D (which we call 'origin deviation') before automatic alignment.

ii) Then we apply the proposed automatic alignment method to all palmprint images to recalculate the direction deviation and origin deviation after the alignment.

iii) Last, we make a comparison of both direction deviation and origin deviation before and after the alignment. Table 3-2 shows that before the alignment the direction deviation is 7.87° and the origin deviation is around 27.87 pixels, but after the alignment the direction deviation becomes 2.07° and the origin deviation is only about 3.17 pixels.

It is obvious that our alignment method can automatically put all the palmprints into their closed locations and directions, which are acceptable in palmprint authentication. Some typical examples before and after the automatic alignment are shown in Figure 3-11.

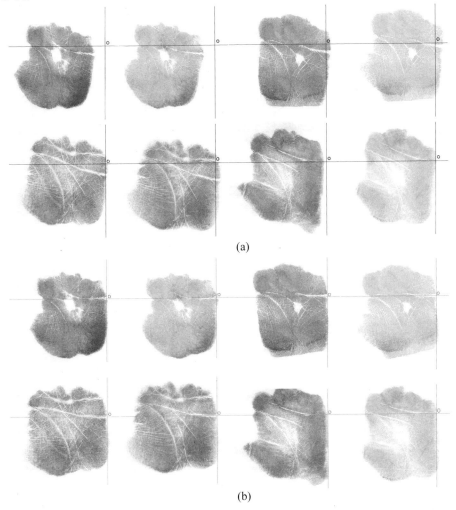

(a)

(b)

Figure 3-11. Experimental results by a group of palmprints, (a) before, and (b) after alignment.

Table 3-2. Average distance comparison between palmprints before and after alignment.

	Origin deviation	Direction deviation
Before alignment	27.87 (pixels)	7.87°
After alignment	3.17 (pixels)	2.07°

3.5   Summary

Palmprint, a new biometric, is regarded as one of the most unique, reliable and stable personal characteristics. Offline palmprint images are collected by inking a user's palm and pressing it onto a sheet of white paper. After the ink has dried, the palmprint image on the paper is digitized by a scanner and stored in the personal computer. In palmprint authentication, the first and crucial step is palmprint segmentation and distortion correction. This chapter first defines the datum points appeared in a palmprint, then reports the method of datum point determination for the palmprint alignment. From the experiments, 286 out of 300 palmprints are found to be in excellent agreement with the manual estimate, which states that this method is acceptable for the palmprint alignment. At last, another new palmprint alignment method, for dealing with image rotation and translation, is proposed. It uses two invariant characteristics of a palmprint, i.e. key point and outer boundary, to automatically put all the palmprints into their closed location and direction. Based on this alignment method, up to 13% correct identification rate can be improved. The experiments illustrate the effectiveness of this method on the palmprint alignment.

4 FEATURE EXTRACTION AND MATCHING

After the preprocessing of the inked palmprint image, as reported in Chapter Three, palmprint authentication requires two further operations: feature extraction and matching. In this chapter, we report a line based feature extraction and matching method. Based on our observations, the curvature of the principal lines is small enough to be represented by several straight line segments. Straight lines are easy to represent and match. In the following discussion, line features are regarded as straight line segments. Section 4.2 describes palmprint matching. Section 4.3 discloses a palmprint classification method based on the singular point. Section 4.4 summarizes this chapter.

4.1 Line Feature Extraction

Line feature extraction is always an important but difficult step in image verification and many line feature detection methods have been proposed [37-39]. Most of these methods cannot generate a precise line map of stripe images such as palmprint images. Although the algorithm of pyramid edge detection based on stack filter performs well for these types of lines [37], it can only extract the principal lines from a palmprint image.

In addition, a 3×3 operator, h_v, which convolves the image with

$$h_v = \begin{vmatrix} -\frac{1}{2} & 1 & -\frac{1}{2} \\ -\frac{1}{2} & 1 & -\frac{1}{2} \\ -\frac{1}{2} & 1 & -\frac{1}{2} \end{vmatrix} \tag{4-1}$$

is commonly used as a detector for thin vertical lines and it can be extended to detect both lines in directions other than the vertical and thick lines [40]. A straight line that passes through the center of a 3×3 neighbourhood must intersect that neighbourhood in one of the following 12 patterns (the center of the neighbourhood has been underlined):

$$\begin{vmatrix} 0 & \theta & 0 \\ 0 & \underline{\theta} & 0 \\ 0 & \theta & 0 \end{vmatrix} \quad \begin{vmatrix} 0 & \theta & 0 \\ 0 & \underline{\theta} & 0 \\ 0 & 0 & \theta \end{vmatrix} \quad \begin{vmatrix} \theta & 0 & 0 \\ 0 & \underline{\theta} & 0 \\ 0 & \theta & 0 \end{vmatrix} \quad \begin{vmatrix} 0 & \theta & 0 \\ 0 & \underline{\theta} & 0 \\ \theta & 0 & 0 \end{vmatrix} \quad \begin{vmatrix} 0 & 0 & \theta \\ 0 & \underline{\theta} & 0 \\ 0 & \theta & 0 \end{vmatrix} \quad \begin{vmatrix} \theta & 0 & 0 \\ 0 & \underline{\theta} & 0 \\ 0 & 0 & \theta \end{vmatrix}$$

$$\quad\;\;(a)\qquad\qquad (b)\qquad\qquad (c)\qquad\qquad (d)\qquad\qquad (e)\qquad\qquad (f)\qquad (4-2)$$

$$\begin{vmatrix} 0 & 0 & \theta \\ 0 & \theta & 0 \\ \theta & 0 & 0 \end{vmatrix} \quad \begin{vmatrix} 0 & 0 & 0 \\ \theta & \theta & \theta \\ 0 & 0 & 0 \end{vmatrix} \quad \begin{vmatrix} 0 & 0 & 0 \\ \theta & \theta & 0 \\ 0 & 0 & \theta \end{vmatrix} \quad \begin{vmatrix} \theta & 0 & 0 \\ 0 & \theta & \theta \\ 0 & 0 & 0 \end{vmatrix} \quad \begin{vmatrix} 0 & 0 & \theta \\ \theta & \theta & 0 \\ 0 & 0 & 0 \end{vmatrix} \quad \begin{vmatrix} 0 & 0 & 0 \\ 0 & \theta & \theta \\ \theta & 0 & 0 \end{vmatrix}$$

    (g)   (h)   (j)   (k)   (l)   (m)

To obtain a complete set of thick-line operators, we would need an analogue of h_v for each of these patterns. For the five near-vertical patterns, then h operator would be defined by comparing the three "θ" points with their horizontal neighbours; for the five near-horizontal patterns, we would compare the points with their vertical neighbours; and for the two diagonal patterns, we could do either type of comparison. A similar set of operators could be used for streak detection, with the θ's representing adjacent nonoverlapping averages rather than individual points. Nevertheless, many ridges and fine wrinkles have the same width as coarse wrinkles except that they are shorter. As a result, the algorithm mentioned above is also difficult to acquire the line features from a palmprint because a mass of ridges and fine wrinkles could dirty the line features.

An improved template algorithm is proposed using four directional templates for line segments determination, as shown in Figure 4-1. It is able to extract the line segments from a palmprint effectively, and then post-process those line segments of each orientation respectively and then combine them. The improved template algorithm consists of the following main steps:

1) Determine vertical line segments by using the five near-vertical templates shown in Figure 4-1 (a), and then thin and post-process these line segments. The rule of post-processing is to clear up the short segments.
2) Similarly, detect lines in other three directions.
3) Combine the results of four different directions, and then post-process once more. The purpose of post-processing is to eliminate the segments overlapped.

Both the process and result of line feature extraction by the improved template algorithm are shown in Figure 4-2.

```
-1 -1 2  2  -1 -1      -1 -1 -1 -1 -1      -1 -1 2  2  -1 -1                           -1 -1 2  2  -1 -1
-1 -1 2  2  -1 -1      -1 -1 -1 -1 -1         -1 -1 2  2  -1 -1                         -1 -1 2  2  -1 -1
-1 -1 2  2  -1 -1       2  2  2  2  2             -1 -1 2  2  -1 -1                   -1 -1 2  2  -1 -1
-1 -1 2  2  -1 -1       2  2  2  2  2                 -1 -1 2  2  -1 -1           -1 -1 2  2  -1 -1
-1 -1 2  2  -1 -1      -1 -1 -1 -1 -1                     -1 -1 2  2  -1 -1   -1 -1 2  2  -1 -1
                      -1 -1 -1 -1 -1
       (a)                  (b)                        (c)                            (d)
```

Figure 4-1. Four improved directional templates for line segments determination: (a) vertical; (b) horizontal; (c) left diagonal; and (d) right diagonal.

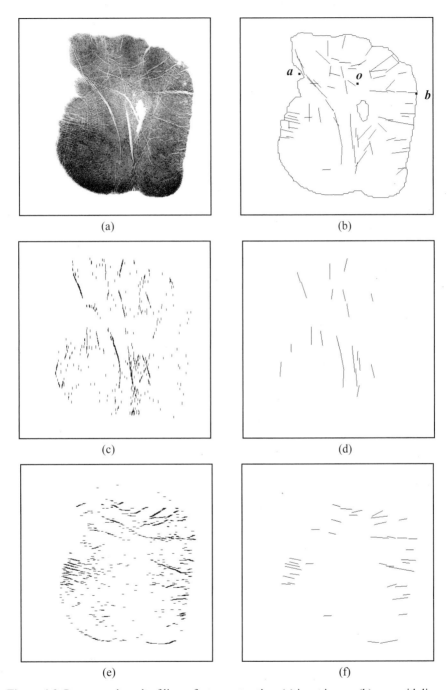

Figure 4-2. Process and result of linear feature extraction: (a) input image; (b) map with linear features extracted; (c), (e), (g) and (i) line segments detection in four different directions by template algorithm, respectively; (d), (f), (h) and (j) their results thinned and post-processed.

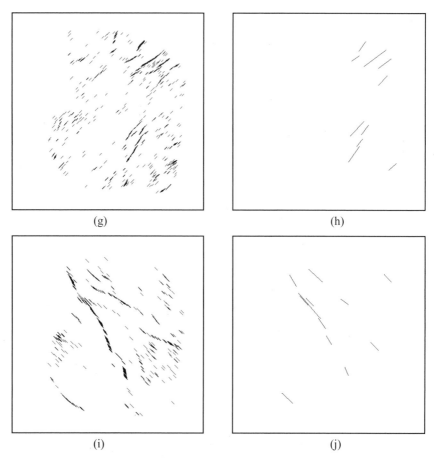

Figure 4-2. (Continued) Process and result of linear feature extraction: (a) input image; (b) map with linear features extracted; (c), (e), (g) and (i) line segments detection in four different directions by template algorithm, respectively; (d), (f), (h) and (j) their results thinned and post-processed.

4.2   Line Matching

In general, there are many ways to represent a line. One way, which is always possible, is to store the endpoints of every straight line segment [41]. In a two-dimensional right angle coordinate system which is uniquely established by the datum points (see Figure 3-2 of Chapter 3), line segments can be described by endpoints: $(X_1(i), Y_1(i))$, $(X_2(i), Y_2(i))$, $i = 1, ..., I$, where I is the number of line segments. Without loss of generality, exchange the endpoints of each line segment so that $X_1(i) \leq X_2(i)$, $i = 1, ..., I$. If $X_1(i) = X_2(i)$, exchange the endpoints so that $Y_1(i) \leq Y_2(i)$. Next, three parameters of each line segment, including slope, intercept and angle of inclination, can be calculated as follows:

$$slope(i) = (Y_2(i) - Y_1(i)) / (X_2(i) - X_1(i)), \tag{4-3}$$

$$intercept(i) = Y_1(i) - X_1(i) \times slope(i) \tag{4-4}$$

and

$$\alpha(i) = \tan^{-1}(slope(i)). \tag{4-5}$$

The object of matching is to tell whether two line segments from a couple of palmprint images are the same in a palmprint. The two-dimensional right angle coordinate system established by the datum points does act as the important registration in line feature matching. For example, two line segments from two images can be represented as $(X_1(i), Y_1(i))$, $(X_2(i), Y_2(i))$ and $(X_1(j), Y_1(j))$, $(X_2(j), Y_2(j))$, respectively. And the Euclidean distances between the endpoints of two line segments are denoted as

$$\nabla_1 = \sqrt{((X_1(i) - X_1(j))^2 + ((Y_1(i) - Y_1(j))^2}, \tag{4-6}$$

$$\nabla_2 = \sqrt{((X_2(i) - X_2(j))^2 + ((Y_2(i) - Y_2(j))^2}. \tag{4-7}$$

Without question, the following conditions for line segment matching can be proposed:

(1) If both ∇_1 and ∇_2 are less than some threshold D, then it clearly indicates that two line segments are same,

(2) If the difference of angle of inclination (difference between two line segments) is less than some threshold β and that of intercepts is less than some threshold B, then it shows that two line segments have the equal angle of inclination and intercept. Within class of equal angle of inclination and equal intercept, if one of ∇_1 and ∇_2 is less than D, then two line segments clearly belong to the same one, and

(3) While two line segments overlap, they are regarded as one line segment if the midpoint of one line segment is between two endpoints of another line.

Verification Function
By applying the above rules of line feature matching to a couple of palmprint images, we can achieve the corresponding pairs of lines (see Figure 4-3). The verification function, r, is defined as

$$r = 2N / (N_1 + N_2), \tag{4-8}$$

where N is the number of these corresponding pairs; N_1 and N_2 are the numbers of the line segments determined from two palmprint images, respectively. In principle, it shows that two images are from one palm if r is more than some threshold T which is between 0 and 1.

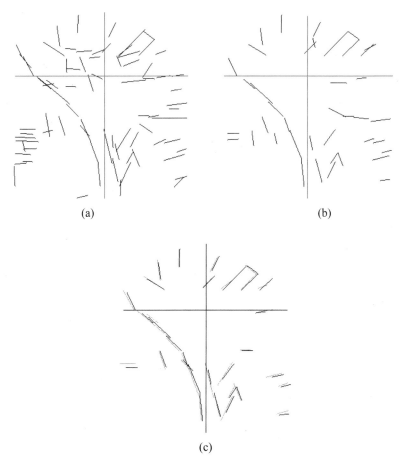

(a)   (b)

(c)

Figure 4-3. Results of a pair of palmprint matching by using line features: (a) template linear
feature set; (b) input linear feature set; and (c) matching result.

Experimental Results

In our experiments, we employ 200 palmprint images from 20 right palms using the
offline palmprints database, **PolyU-OFFLINE-Palmprint-I**. Firstly, we use the
algorithm of datum point determination (described in Section 3.2 of Chapter 3) to
preprocess all the palmprint images for the palmprint registration. Then, the
palmprint verification with both datum point invariance and line feature matching has
been tested. A slippery pair of statistics known as false rejection rate (FRR) and false
acceptance rate (FAR) represents the measure of experimental result. In this
experiment, some thresholds are adopted as $D = 5$, $\beta = 0.157$ and $B = 10$. The results
with various thresholds, T, are shown in Figure 4-4 and the palmprint verification can
obtain an ideal result while T is between 0.1 and 0.12.

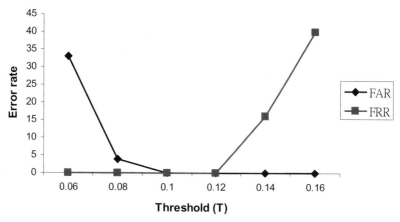

Figure 4-4. Experimental results for FAR and FRR ($D = 5$, $\beta = 0.157$ and $B = 10$).

4.3 Singular Point for Classification

All palmprints are made up of both ridges and valleys, similar to that of fingerprint, where their flow characteristics are quite regular. There are some classes of ridges in a palmprint, such as loop, whorl and arch formed by ridges (see Figure 4-5). Singular point [35] is defined as the core and delta points formed by the ridge pattern of a palmprint. Delta point is defined as the point of bifurcation (in a delta-like region) on a ridge splitting into two branches which extend to compass the complete pattern area. Notice that all delta points are born features of everyone as they are composed of the ridges. That means singular points, including core and delta points appeared in the palmprint have the stability and uniqueness for the palmprint classification.

Singular Point Detection
A point in the directional image is classified as an ordinary, core or delta point by computing the field of the Poincaré Index (PI) along a small closed curve around the point [23]. We can obtain the PI by summing up the changes in the direction angle around the curve. It shows that the PI of delta point is positive and that of core point is negative when a full counter-clockwise turn is made around the curve in a directional image. Here we use the symbols, "—", "|", "\" and "/", to represent four different directions, respectively. We propose a simplified algorithm to detect these singular points as follows:

(i) Generate a four-directional image. Each palmprint can be described by a directional image which represents the flow characteristics of ridges. Four directions are defined as vertical, horizontal, left diagonal and right diagonal. [36] These directions can be represented by 90, 0, 135 and 45 degrees, respectively.

(ii) Detect singular points. In the directional image, a pixel whose eight neighborhoods have more than two directions can be confirmed as a singular point. According to their PIs, core and delta points are then distinguished.

(iii) Eliminate the false singular points. Based our experiments, some rules to delete the false points can be given. If a pair of core points (or delta points) is too close, then they are considered as one core point (or delta point); if a core point nears to a delta point, all of them are taken away.

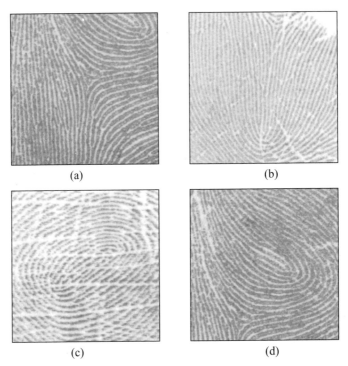

(a) (b)

(c) (d)

Figure 4-5. Classes of ridges in palmprint: (a) delta point, (b) arch, (c) whorl, and (d) loop.

As a result, Figure 4-6 shows that two singular points, core and delta, are represented in an original image and its directional image, respectively. It is not doubtful that the singular points in the directional image can be simply found out.

Outside Region Classification

We always can get a complete outside region from an inked palmprint image. There are abundance shapes consisting of the singular points at the outside region. From hundreds of palmprint images, we summarize fifteen different distribution maps based on the number and their locations of singular points (see Figure 4-7).

For simplicity, we only use the number of singular points at the outside region as known criterion to classify palmprint images. It shows that there are six typical cases given in our palmprint classification, including (*a*) no singular point; (*b*) one delta point; (*c*) one core point; (*d*) one core point and delta point; (*e*) two delta points and one core point; and (*f*) others. Six typical cases of singular points in the outside region for palmprint classification, both original and directional images, are demonstrated in Figure 4-8.

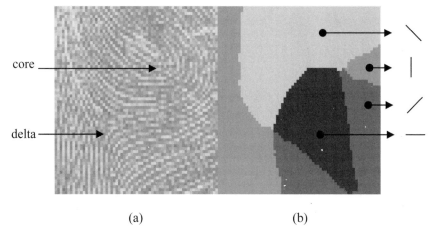

core

delta

(a)   (b)

Figure 4-6. Singular points in both original image and directional image: (a) original image; (b) directional image.

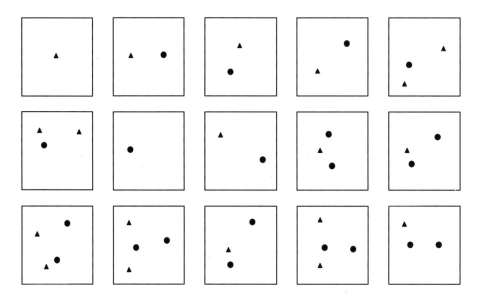

Figure 4-7. Typical contribution map of singular points at outside region of palmprint ("Δ" - delta point and "•" - core point)

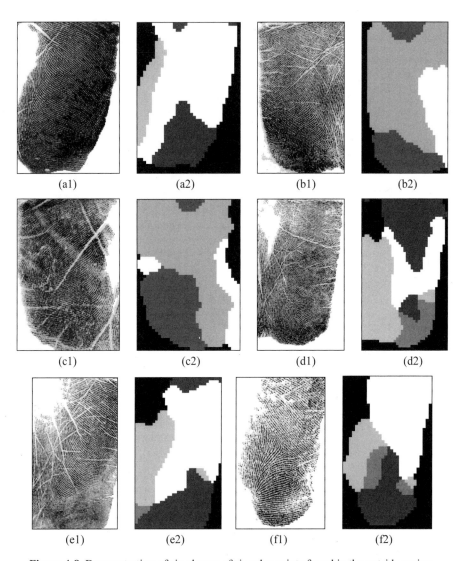

Figure 4-8. Demonstration of six classes of singular points found in the outside region, original image: (a1)-(f1); directional image: (a2)-(f2).

Experimental Results

In order to test the classification method based on singular points, we picked 354 palmprint images from the database, **PolyU-OFFLINE-Palmprint-I**, in which the outside region are extracted to classify six classes of palmprints. Table 4-1 lists the classification results, in which around 97% correct rate can be obtained. It demonstrates that this method is acceptable, which can speed up the palmprint authentication in a large database (i.e. millions of palmprints).

Table 4-1. Classification results in outside region.

Assigned class	True class					
	a	*b*	*c*	*d*	*e*	*f*
a	**132**	4	1	2	0	0
b	0	**149**	0	0	0	2
c	0	0	**15**	1	0	0
d	0	0	0	**27**	1	0
e	0	0	0	0	**8**	0
f	0	0	0	0	0	**12**

4.4 Summary

This chapter revealed a line based feature extraction and matching strategy from the offline palmprint images. Obviously, line feature includes curves and straight lines. Based on our observation, the curvature of the principal lines is small enough to be represented by several straight line segments. Straight lines are easy to represent and match. The matching of palm lines can be performed by measuring the Euclidean distances between the endpoints of two line segments and the angles between these line segments. Experimental results show that this method is effective.

In addition, a classification method is proposed based on the singular points, such as loop, whorl and arch formed by ridges, found on the outside region of a palm. There are six typical cases of singular points in the outside region for palmprint classification. Experimental results show that around 97% correct classification can be achieved in a database of 354 offline palmprint images.

5 LAYERED IDENTIFICATION USING MULTIPLE FEATURES

This chapter describes a new method to authenticate individuals using offline palmprints with multiple features, searched in a layered fashion. We first introduce the multiple features of a palmprint such as texture features and interesting points. We then propose a dynamic texture-based selection scheme to facilitate fast hierarchical searches for the best match of the sample against the database. Section 5.2 describes global texture energy, which is used to guide the dynamic selection of a small set of similar candidates from the database at a coarse level for further processing. Section 5.3 introduces an interesting-point-based image matching that is performed on the selected similar patterns at fine level for final confirmation. Section 5.4 provides experimental results and Section 5.5 summarizes this chapter.

5.1 Introduction

A key issue in palmprint identification involves the search for the best matching of the input from the templates in the palmprint database. The selection of features and similarity measures are two fundamental problems to be solved. A feature with good discriminating ability should exhibit a large variance between individuals and small variance between samples from the same person. Principal lines and datum points are regarded as useful palmprint features [42] and have been successfully used for verification. In addition, there are many other features associated with a palmprint [43], such as geometry features, wrinkle features, delta point features and minutiae features, revealed in Chapter 3.

It is noted that all of these features are concerned with the local attributes based on points or line segments. The lack of global feature representation resulted in the high computation demanded for matching which measures the degree of similarity between two sample sets. Although a line feature matching method is reported in Chapter 4, to be powerful for easy computation, tolerance to noise and high accuracy in palmprint verification by using both line features and datum points [18, 44], it has the following limitations: 1) for a given test sample, the line matching process was applied to all of the template sets to search for the best matching. For a large collection of templates, the computational burden is high; 2) for an occluded palmprint sample, the principal lines and endpoints might be missing, which would cause the consequent failure of the line matching algorithm. In addition, the traditional exhaustive comparison methods are very time consuming for a large palmprint database of palmprints.

In view of this situation, we proposed a dynamic selection scheme to facilitate the coarse-to-fine palmprint pattern matching by combining global and local palmprint features in a hierarchical fashion. The global texture energy (GTE) is introduced to represent the global palmprint feature, which is characterized with high convergence of inner-palm similarities and good dispersion of inter-palm discrimination. Such a global feature is used to guide the dynamic selection of a small set of similar candidates from the database at coarse level for further matching. Interesting points are then used as the local features for the matching in the final stage to produce a matching result.

5.2   Coarse-level Classification

Palmprints consist of large numbers of thin and short line segments represented in forms of wrinkles and ridges. Such patterns can be well characterized by texture. The global texture energy is used to represent the global feature of a palmprint. Then, a texture-based dynamic selection scheme for palmprint identification is performed at the coarse level.

Texture Feature Measurement

We calculate the global texture energy and let it directly function as a classifier. Such a statistical approach proposed by Laws [45] is notable for its computational simplicity.

Let $I_{M \times N}$ denotes a palmprint image with size of $M \times N$, $I(i, j)$ denotes a point on image $I_{M \times N}$. Global texture energy of $I_{M \times N}$ is denoted as GTE_I. Then GTE_I is defined as:

$$GTE_I = \frac{\sum_{i=1}^{M} \sum_{j=1}^{N} E(i, j)}{M \times N} \tag{5-1}$$

$$E(i, j) = \frac{1}{(2n+1)^2} \sum_{k=i-n}^{i+n} \sum_{l=j-n}^{j+n} |F(k, l)| \tag{5-2}$$

$$F(i, j) = A(i, j) * I(i, j) \tag{5-3}$$

$$= \sum_{k=-a}^{a} \sum_{l=-a}^{a} A(k, l) I(i+k, j+l) \tag{5-4}$$

where *A(i,j)* is a zero sum mask with size *2a+1* by *2a+1* , in this case, the size of the mask *A(i,j)* is 5×5 and * denotes a 2D convolution. *E(i,j)* is calculated through *F(i,j)* within a *2n+1* by *2n+1* window at point *(i,j)* , in this case, it is 15×15. For the purpose of fast selection of a small set of similar palmprint patterns from the database, we use four 'tuned' masks to capture the global palmprint texture features which are more sensitive to horizontal lines, vertical lines, 45° lines and -45° lines respectively. Figure 5-1 lists the four of our modified 'tuned' masks. Such a texture energy measurement for global palmprint feature extraction has the following characteristics:

- insensitive to noise,
- insensitive to translation,
- easy to compute,
- high convergence within the group and good dispersion between groups.

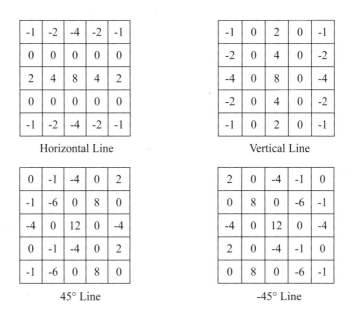

Figure 5-1. The four 'tuned' masks used in global palmprint texture feature extraction

Similar Palmprint Pattern Grouping

It is very important to group similar palmprint patterns from the given palmprint database for further identification and verification. This grouping task can be viewed as a decision-making process which allocates an input palmprint sample to those categories with the similar measurements in the database. The global texture energy exhibits high convergence of inner-palm similarities and good dispersion of inter-palm discrimination.

Figure 5-2 shows four palmprint samples from four different individuals with distinctive texture features and Figure 5-3 demonstrates the distribution of global palmprint texture energy measurements.

Although different individuals have different palmprint patterns, some of these patterns are so similar that it is very difficult, if not impossible, to classify them based on the global texture features alone. Figure 5-4 shows samples of similar palmprint patterns from different groups. To tackle such a problem, we propose a dynamic selection scheme to group a small set of the most similar candidates in the database for further identification by image matching at fine level. The idea behind this is to eliminate those candidates with large GTE differences and generate a list of the very similar candidates with very close GTEs. The following summarizes the main steps for implementation:

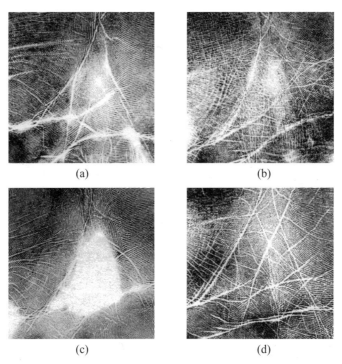

(a)   (b)

(c)   (d)

Figure 5-2. Samples of different palmprint patterns with distinctive texture features. (a) strong principle lines, (b) full of wrinkles, (c) less wrinkles, and (d) strong wrinkles.

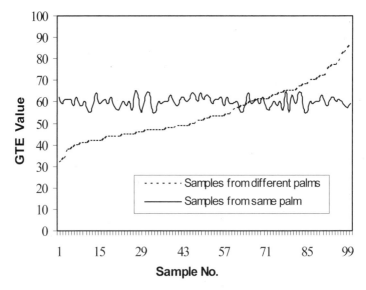

Figure 5-3. The comparison of palmprint GTE distribution: inter-palm dispersion vs. inner-palm convergence.

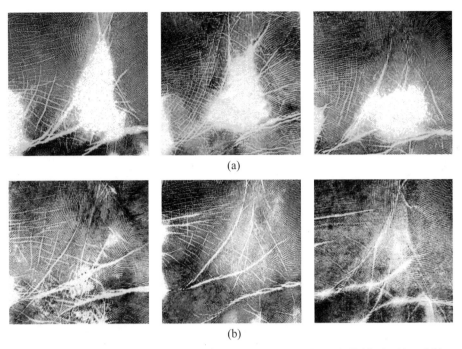

(a)

(b)

Figure 5-4. Samples of the similar palmprint patterns from different individuals: (a) and (b).

Candidates Selection by GTE

There are five steps for selecting the candidates by GTE as follows:

- Step 1: Convolve the sample palmprint image I_{sample} with the four tuned masks A_i, $i=1,2,3,4$ and obtain the corresponding global texture energy $GTE_{sample}(i)$, $i=1,2,3,4$.

- Step 2: Compare the sample with the candidate in the database in terms of GTE and calculate their difference d, where d is given by:

$$d = \sum_{i=1}^{4} | GTE_{sample}(i) - GTE_{sample}(i) | \qquad (5\text{-}5)$$

- Step 3: If d is smaller than the pre-defined threshold value, this candidate is added to the list for further matching.

- Step 4: Go to Step 2 and repeat the same procedure until all of the candidates are considered.

- Step 5: Provide the final list of candidates to guide the search for the best matching at fine level.

5.3   Fine-level Identification

A guided image matching scheme is developed to determine the degree of similarities between the template pattern and every possible sample pattern selected throughout the texture classification procedure which is reported in Section 5.2. Unlike the traditional image matching methods which are based on the detection of edge points, we proposed to detect interesting points in textured images to achieve high performance. To avoid the blind searching for the best fit between the given samples, we applied a dynamic selection of interesting points to search for the best matching in a hierarchical structure.

Feature Points – Interesting Points vs. Edge Points

Most matching algorithms are based on binary images to identify the interested object(s). Therefore, the original image, either grayscale or color images, should be converted into a binary image. Traditional methods that convert an original image into a binary image rely on edge detection. Even though edge detection has been successfully used for many years mostly due to its simplicity, it has some problems which prevent it from being applied on a real-time image matching scheme, such as

- it is susceptible to noise in the image.
- feature points may not be well distributed.
- it produces a large number of feature points for an image, many of the points however are redundant.

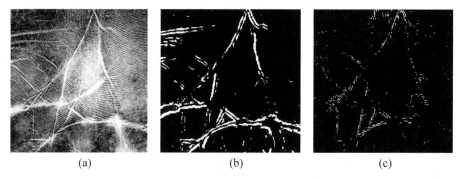

(a)   (b)   (c)

Figure 5-5. Comparison of palmprint feature point detection: edge vs interesting points. (a) histogram equalized original palmprint, (b) edge detected, and (c) interesting points.

The redundant information limits the possible speedup in subsequent operations since the time consumption of a matching algorithm, such as Borgefors' hierarchical chamfer matching [46], and the Hausdorff distance [47], is directly related to the number of feature points used in matching. The more the points are, the more the time is required. Therefore, it is desirable to use an algorithm which only extracts those feature points that

- are representative and distinctive, such as corners, and without redundancy.
- are robust to "noise".

This has prompted the research in [48] to use interesting point detectors rather than edge detectors to extract feature points from a given image for matching. You *et al.* [48] indicates the Plessey operator is superior to the Moravec operator according to the criteria set above. Hence, the Plessey operator is chosen as the interesting point detector in this work. Figure 5-5 shows the comparison of edge points and interesting points in representing the original image. Figure 5-5 (a) is a histogram equalized image of size 232 × 232 ranging from 0 - 255 in grayscale, Figure 5-5 (b) shows edge points detected by Prewitt operator and Figure 5-5 (c) shows interesting points detected by Plessey operator.

Distance Measurement for Image Matching
Given two images, a matching algorithm determines the location of the template image on the target image and places a value on their similarity at this point. The value determines the degree of similarity. Based on Huttenlocker *et al.* [47], we are able to use a Hausdorff distance algorithm to search for portions, or partial hidden objects. This feature also allows us to partition the target image into a number of sub images, and to then carry out the matching process simultaneously on these sub images in order to accelerate the process. Instead of using edge points, we use interesting points as a basis for computing the Hausdorff distance. The objective is to reduce the computation required for a reliable match.

The hierarchical guided matching scheme was first introduced by Borgefors [46] in order to reduce the computation cost required to match two images. We extended it by using interesting points rather than edge points in a similar fashion, i.e., an interesting point pyramid is created and the matching starts from the lowest resolution and the results of this match guides the search on the possible area of the higher resolutions. We further extend it by using the Hausdorff distance as a measure of similarity. The advantage of using Hausdorff distance in a matching process relies on the capability of searching for portions of images, which allows us to partition the target image into a number of sub images and simultaneously match the template image on these sub images.

The Hausdorff distance is a non-linear operator which measures the mismatch of the two sets. In other words, such a distance determines the degree of the mismatch between a model and an object by measuring the distance of the point of a model that is farthest from any point of an object and vice versa. Therefore, it can be used for object recognition by comparing two images which are superimposed on one another. The key points regarding this technique are summarized as below.
Given two finite point sets $A = \{a_1,...,a_m\}$ and $B = \{b_1,...,b_n\}$, the Hausdorff distance D_H between these two sets is defined as

$$D_H = \max(d_{AB}, d_{BA}) \tag{5-6}$$

where d_{AB} is the distance from set A to set B expressed as

$$d_{AB} = \max_{a_i \in A}(d_{a_i B}) \tag{5-7}$$

while $d_{a_i B}$ is the distance from point a_i to set B given by

$$d_{a_i B} = \min_{b_j \in B}(d_{a_i b_j}) \tag{5-8}$$

Obviously the Hausdorff distance D_H is the maximum of d_{BA} and d_{BA} which measures the degree of mismatch between two sets A and B.

In general, image data are derived from a raster device and represented by grid points as pixels. For a feature detected image, the characteristic function of the set A and B can be represented by a binary array $A[i,j]$ and $B[i,j]$ respectively, where the (i, *j)th* entry in the array is non-zero for the corresponding feature pixel in the given image. Therefore, distance array *D[i,j]* and *D'[i,j]* are used to specify for each pixel location *(i,j)* the distance to the nearest non-zero pixel of *A* or *B* respectively, where *D[i,j]* denotes the distance transform of *A* and *D'[i,j]* denotes the distance transform of *B*. Consequently, the Hausdorff distance as a function of translation can be determined by computing the point-wise maximum of all the translated *D* and *D'* array in the form of:

$$F(i, j) = \max(\max_a, \max_b) \qquad (5\text{-}9)$$

where

$$\max_a = \max_a D[a_i - i, a_j - j] \qquad (5\text{-}10)$$

$$\max_b = \max_b D'[b_i + i, b_j + j] \qquad (5\text{-}11)$$

Layered Matching Scheme

The hierarchical image matching scheme was first proposed by Borgefors [46] in order to reduce the computation required to match two images. This section details our extension of this scheme by introducing a guided search strategy to avoid the blind searching for the best fit between the given patterns. In order to avoid the blind searching for the best fit between the given patterns, a guided search strategy is essential to reduce computation burden. Our extension of the hierarchical image matching scheme (HIMS) was based on a guided searching algorithm that searches first at the low level, coarse grained images, to the high level, fine grained images. To do this we needed to obtain a Hausdorff distance approximation for each possible window combination of the template and target image at the lowest resolution. Those that returned a Hausdorff distance approximation equal to the lowest Hausdorff distance for those images were investigated at the higher resolution. The following summarizes the key steps involved in a HIMS algorithm.

Hierarchical Image Matching Scheme (HIMS)

There are 4 steps for the HIMS:

- Step 1   create image pyramid
- Step 2   for all combinations of windows
                at lowest level
                get value of match for this combination
                if low value add to lowest list
            end-for
- Step 3   for each remaining level
                remove area from lowest list
                get match value for this area
                if low value add to lowest list
            end-for

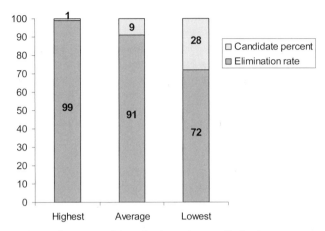

Figure 5-6. The performance of the selection scheme: elimination rate vs. candidate percentage

5.4 Experimental Results

The palmprint image samples used for the testing are obtained from the offline palmprints database, **PolyU-OFFLINE-Palmprint-II**, by first extracting the subimage (of size 232 × 232) from the original one. A series of experiments have been carried out to evaluate the performance of the proposed algorithms from the extracted sub-image (200 palmprint sub-images totally).

Laws' texture energy concept is extended by the use of four "tuned" masks to extract the global palmprint texture features sensitive to the horizontal lines, vertical lines, 45° lines and -45° lines. Such a global texture energy measurement is used to guide the fast selection of the small set of the most similar palmprint patterns for fine matching. For a given test palmprint sample, on the average, 91% of the candidates in the database are classified as distinctive from the input data and filtered out at the coarse classification stage. In the worst case, the elimination rate of the candidates is 72% and only 28% of the samples are remained for further identification at fine level by image matching. Figure 5-6 demonstrates the effectiveness of this selection scheme by global palmprint texture feature extraction.

An interesting point based image matching is performed for the final confirmation at fine level by the proposed hierarchical structure. The average accuracy rate is 95%. Since the majority samples have been filtered out from by the coarse classification, the execution speed of the fine level feature matching has been increased significantly.

5.5 Summary

Palmprint feature extraction and matching are two key issues in palmprint identification and verification. In contrast to the traditional techniques, this chapter proposed a dynamic selection scheme that measures global texture features and detects local interesting points. Our comparative study of palmprint feature extraction shows that palmprint patterns can be well described by textures, and the texture energy measurement possesses a large variance between different classes while remaining high compactness within the class. The coarse-level classification using global texture features is effective and essential to reduce the number of samples for further processing at fine level. Layered searching scheme based on the interesting points improves the system efficiency further. The use of Hausdorff distance as the matching criterion can handle the partial occluded palmprint patterns, which makes our algorithm more robust. The experimental results provide the basis for the further development of a palmprint authenticated system based on a large database, with high performance in terms of effectiveness, accuracy, robustness and efficiency.

PART III

ONLINE METHODOLOGIES

6 PALMPRINT SEGMENTATION BY KEY POINT FEATURES

Palmprint segmentation is the process of correcting distortions and putting all palmprints under the same coordinate system so that the expected area of each palmprint can be extracted for feature extraction and matching. Inked palmprints (offline) identification method was introduced in [18], where two key points in a palmprint were defined and determined automatically by combining projection and line tracing methods. Since the image quality of inked palmprints is different from that of online captured images, the offline segmentation method introduced in [18] is not applicable to online palmprint segmentation. This chapter reports two palmprint segmentation approaches based on online palmprints. Section 6.1 describes square-based approaches that can be used to extract the region of interest (ROI) called the *central part sub-image* of a palmprint. Two different square-base approaches are tested on the **PolyU-ONLINE-Palmprint-I** database and **PolyU-ONLINE-Palmprint-II** database. Section 6.2 illustrates an inscribed circle-based approach for the central part palmprint sub-image. Section 6.3 compares the square-based and circle-based approaches to determine the effectiveness of these two approaches. Section 6.4 summarizes the chapter.

6.1 Square-Based Segmentation

Way 1: PolyU-Online-Palmprint-I
The main idea of square-based segmentation approach is to first define three key points on a palmprint. An orthogonal coordinate system is then given by using these three points [49]. During segmentation, a square with a fixed size is extracted from a predefined position under the coordinate system. The size and position of the square are determined by using statistics from many palmprints. The basic rule in determining the size and position of the square is to ensure that the part of the image that is extracted is available in all palmprints, with rich palmprint features. Figure 6-1 shows two palmprints samples in our database, **PolyU-ONLINE-Palmprint-I**. They are placed in different orientation of different palms size.

Definitions of Key Points and the Orthogonal Coordinate System on a Palm
There are three key points on a palm, namely k1, k2, and k3 (see Figure 6-2). k1 is the midpoint between the index finger and middle finger, k2 is the midpoint between the middle finger and the ring finger, and k3 is the midpoint between the ring finger and the little finger. We lined up k1 and k3 to get the Y-axis of the palmprint coordinate system. We then take a line through k2, perpendicular to the Y-axis, to determine the origin of the palmprint coordinate system (see Figure 6-2). Here, we

suppose that the fingers are not stuck together and at least four fingers (index, middle, ring and little finger) are present in an image. Therefore, the three gaps between the fingers are obtained, and the three key points are found.

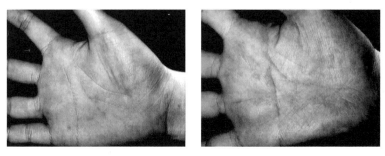

Figure 6-1. Palmprint samples in our database, **PolyU-ONLINE-Palmprint-I.**

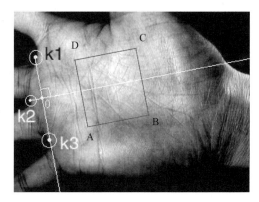

Figure 6-2. Orthogonal coordinate system on a palmprint (k1, k2, k3 are midpoints between two fingers; ABCD is the square in the central part of the palm).

The steps on obtaining the square area ROI, called *central part sub-image* of a palmprint, from these key points are summarized as follows:

1) Use a threshold, α, to convert the original gray image (see Figure 6-3 (a)) into a binary map (see Figure 6-3 (b)).

$$I_{BinaryMap}(i,j) = \begin{cases} 1 & I_{GrayMap}(i,j) \geq \alpha \\ 0 & I_{GrayMap}(i,j) < \alpha \end{cases} \tag{6-1}$$

2) Smooth the binary map by a Gaussian filter in Figure 6-3 (c).

$$I_{SmoothedMap} = I_{BinaryMap} * A \tag{6-2}$$

where A is the Gaussian filter.

3) Trace the boundary of the holes between the fingers as shown in Figure 6-3 (d).

4) Calculate the center of gravity of the holes and decide the key points – k1, k2, and k3, respectively, as shown in Figure 6-3 (e).

5) Line up k1 and k3 to get the Y-axis of the palmprint coordinate system and then make a line through k2, perpendicular to the Y-axis to determine the origin of the palmprint coordinate system, as shown in Figure 6-3 (f).

6) Once the coordinate system is decided, a fixed size sub-image of the central part of a palmprint is extracted. In Figure 6-3 (f), the square "*ABCD*" is the area that is extracted. The position of "*A*" in the coordinate system and the size of "*ABCD*" are fixed in all palmprints. Figures 6-4 show some central part palmprint sub-image extracted using this approach.

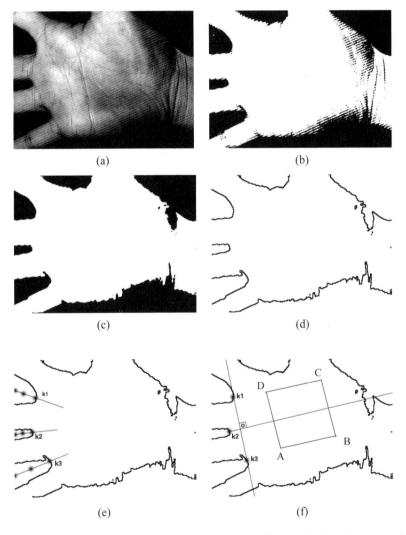

Figure 6-3. Determination of k1, k2 and k3: (a) the original image, (b) the binary map of the original image, (c) smoothed map of (b), (d) palmprint boundary, (e) determination of k1, k2 and k3, and (f) determination of the coordinate system.

This segmentation approach needs the accurate determination of the three holes between the fingers. Otherwise, it has the possibility of getting a wrong central part sub-image, and finally affects the accuracy of the palmprint identification.

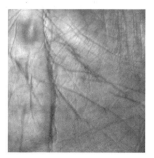 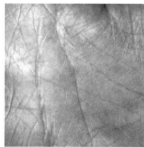 

Figure 6-4. Palmprint central part sub-image is extracted.

Way 2: PolyU-Online-Palmprint-II

This palmprint preprocessing approach uses the holes between fingers as the parameters to build a coordinate system for aligning different palmprint images [50]. This preprocessing defines the central part palmprint sub-image for feature extraction. Figure 6-5 shows some palmprints samples in our database, **PolyU-ONLINE-Palmprint-II**.

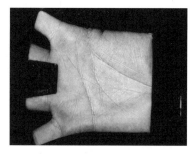 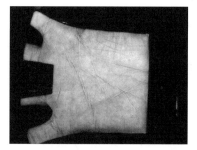

Figure 6-5. Palmprint samples in our database, **PolyU-ONLINE-Palmprint-II**.

There are five major steps for the palmprint segmentation (see Figure 6-6):

Step 1: Binary the image
Apply a lowpass filter, L(u, v), such as Gaussian, to an original image, O(x, y). Then, a threshold, T_p, is used to convert the convoluted image to a binary image, B(x, y), as shown in Figure 6-6 (b). Mathematically, this transformation can be represented as

$$B(x, y)=1 \text{ if } O(x,y) * L(x,y) \geq T_p, \tag{6-3}$$

$$B(x, y)=0 \text{ if } O(x,y) * L(x,y) < T_p, \tag{6-4}$$

where *B(x,y)* and *O(x,y)* are the binary image and the original image, respectively; *L(x,y)* is a lowpass filter, such as Gaussian, and "*" represents an operator of convolution.

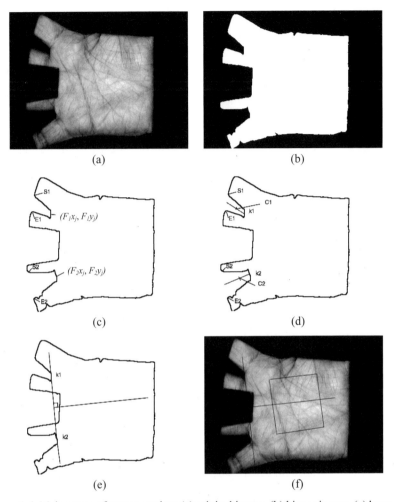

Figure 6-6. Major steps of preprocessing: (a) original image, (b) binary image, (c) boundary tracking, (d) key points (k_1 and k_2) detecting, (e) the coordinate system and (f) the central part of a palmprint.

Step 2: Boundaries tracing
Extract the boundaries of the holes, $(F_i x_j, F_i y_j)$, $(i=1,2)$, between fingers using a boundary-tracking algorithm. The start points, (Sx_i, Sy_i), and end points, (Ex_i, Ey_i), of the holes are then marked in the process (see Figure 6-6 (c)).

Step 3: Compute the center of gravity
Compute the center of gravity, (Cx_i, Cy_i), of each hole with the following equations:

$$Cx_i = \frac{\sum_{j=1}^{M(i)} F_i x_j}{M(i)}, \qquad (6\text{-}5)$$

$$Cy_i = \frac{\sum_{j=1}^{M(i)} F_i y_j}{M(i)},$$

(6-6)

where $M(i)$ represents the number of boundary points in the hole, i. Then construct a line that passes through (Cx_i, Cy_i) and the midpoint of (Sx_i, Sy_i) and (Ex_i, Ey_i). The line equation is defined as:

$$y = x \frac{(Cy_i - My_i)}{(Cx_i - Mx_i)} + \frac{My_i Cx_i - Mx_i Cy_i}{Cx_i - Mx_i},$$

(6-7)

where (Mx_i, My_i) is the midpoint of (Sx_i, Sy_i) and (Ex_i, Ey_i). Based on these lines, two key points, (k_1, k_2), can easily be detected (see Figure 6-6 (d)).

Step 4: Construction of the palmprint coordinate system
Line up k_1 and k_2 to get the Y-axis of the palmprint coordinate system and make a line through their midpoint which is perpendicular to the Y-axis, to determine the origin of the coordinate system (see Figure 6-6 (e)). This coordinate system can align different palmprint images.

Step 5: Extract a central part sub-image
Extract a sub-image with the fixed size on the basis of coordinate system, which is located at the certain part of the palmprint for feature extraction (see Figure 6-6 (f)).

6.2   Inscribed Circle-Based Segmentation

In contrast to the square-based segmentation, this approach is based on an inscribed circle to extract the central part of a palmprint; therefore a round image is obtained rather than a square. Figure 6-7 shows the square and circle area central part subimage on the same palm in our database, **PolyU-ONLINE-Palmprint-I**.

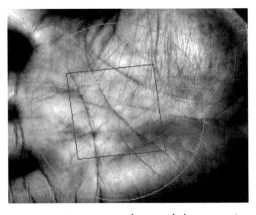

Figure 6-7. Square and circle area central part sub-image on the same palmprint.

The basic idea of using an inscribed circle is to calculate the inscribed circle that meets the boundary of a palm so that it can extract as large an area as possible from the central part of the palmprint (see Figure 6-8).

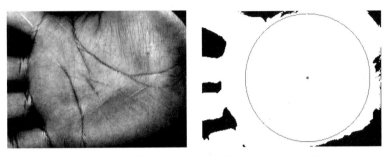

Figure 6-8. The illustration of a circle-based segmentation.

The following describes the details of this approach:

Step 1: Determining the palmprint's boundary
Determining the boundary is a rather simple step. First, we convert the gray scale image into a binary image using the Equation 6-1 and 6-2. Then, we smooth the boundary between the white and black pixels and obtain the contour of a palmprint.

Step 2: Calculation of the inscribed circle
We calculate the biggest inscribed circle for the contour of the palmprint and obtain its center and radius. Because different palms are different sized, the radii may be different. Samples of the same palm have circles with similar radii and centers. When matching two palmprints, we first check the radii. If the radii are not close to each other, we can conclude immediately that the two palmprints are from different palms. Otherwise, further matching is needed within the circle. We suppose that the centers of two palmprints are close but not necessarily exactly at the same point on a palm. Therefore, the matching of palmprints needs to shift and rotate one of the palmprints.

Step 3: Extraction of the central part
Once the circle is determined, the extraction only involves obtaining all pixels inside the circle. Figure 6-9 shows several segmented central part palmprint sub-images.

Figure 6-9. Extracted samples using inscribed circle-based segmentation approach.

6.3 Comparison between two Segmentation Approaches

In this chapter, two kinds of palmprint segmentation approaches based on the online palmprints are reported. A comparison and analysis of the square-based and inscribed circle-based approach is performed to evaluate the effectiveness of these approaches. Three criteria are used in evaluating a segmentation approach, they are: 1) How precisely it extracts the required part? 2) Can the algorithm deal with more distortions in an image? and 3) Does the extracted part include more useful features? By using these criteria, we can obtain the effectiveness of these two palmprint segmentation approaches accordingly.

Accuracy Test

A. Definitions

Commonly, there are three main creases on a palm called principal lines. Here, we defined two key points on a palm according to two principal lines flow out of a palm – A and B in Figure 6-10, respectively. In Figure 6-10, we also define the center of the square as O1 and the center of the circle as O2. In addition, the distance between O1 (O2) and A is denoted as $D_{square-a}$, ($D_{circle-a}$) and the distance between O1 (O2) and B is $D_{square-b}$ ($D_{circle-b}$), respectively.

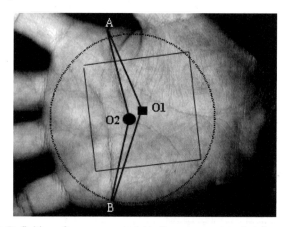

Figure 6-10. Definitions for accuracy test (A, B are two points that flow out from two principal lines; O1 is the center of the square and O2 is the center of the circle).

B. Design of the Experiment

We obtain 400 palmprints from 80 palms, and manually decide the two key points (A,B) of all palms. Then automatically calculate all distances, including $D_{square-a}$, $D_{square-b}$, $D_{circle-a}$ and $D_{circle-b}$. We then generate the experimental results and their comparison.

C. Experimental results

There are five palmprint samples of each palm. It is expected that the extracted parts from each sample include the same area of the palm. We calculate the average distance between the centers of five squares and circles to decide which one moves by the least amount. Our experiment shows that both the shifts of $D_{square-a}$ and $D_{square-b}$ are 6 pixels, and the shifts of $D_{circle-a}$ and $D_{circle-b}$ are 5 pixels, respectively. That is, the circle-based approach is a little more precise than the square-based approach.

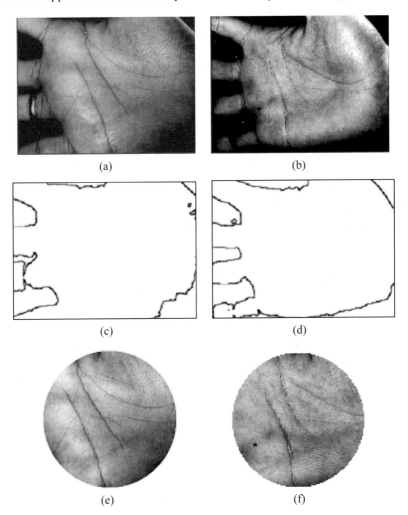

(a)   (b)

(c)   (d)

(e)   (f)

Figure 6-11. Palmprint central part sub-image extraction for special palmprints.

Tolerance to Palmprint Images of Different Quality

In some cases, a palmprint cannot be segmented properly using the square-based approach while it can be processed correctly using the circle-based approach. In our database, there are about 8,000 palmprint images. Only 2/3 of them can be correctly segmented using the square-based approach, but all of them can be segmented correctly using the circle-based approach. Figure 6-11 (a) shows a sample in which the finger has a ring. Figure 6-11 (c) is the binary map of Figure 6-11 (a) that the holes between the fingers cannot be identified correctly due to such a problem. As a result, this palmprint cannot be correctly segmented using the square-based approach. On the other hand, by using the circle-based segmentation approach, the central part palmprint sub-image can be extracted successfully, as shown in Figure 6-11 (e). Another example in Figure 6-11 (b) shows that the little finger cannot be located correctly due to its too dark which can not be separated from the background. Figure 6-11 (d) gives the binary map of Figure 6-11 (b) and it is clear that the hole between ring finger and little finger is not traced correctly. So Figure 6-11 (b) is not satisfactorily segmented using square-based approach. The circle-based approach can segmented it and the round part segment of Figure 6-11 (b) is put there as Figure 6-11 (f).

Comparison of the Sub-image Sizes

Segmentation involves extracting the central part of a palmprint for feature extraction and matching. The size used for square-based approaches are about $128 \times 128 = 16,384$ pixels. On the other hand, the size of a circle varies from palm to palm, which uses the largest inscribed circle that meets the boundary of a palm. The average size of circle from the database is $100 \times 100 \times 3.14 = 31,400$ pixels (i.e., the average radius of the inscribed circles is 100 pixels), therefore, more features can be included. Samples from the same palm may have different circle sizes, but the difference is usually within 2-3 pixels which does not cause a problem for palmprint matching. Different palms may have larger difference with their circle sizes. It provides a useful means on palmprint classification, i.e. use the circle size as the first-step of the classification criteria. If two palmprint samples are too different with circle size, we may conclude immediately that they are from different palms. Only when the circles of two palmprints are of a similar size, they will be compared with the other features in order to decide whether they are from the same palm.

6.4   Summary

Three key issues are involved in palmprint identification: segmentation, feature extraction and feature matching. In this chapter we present two kinds of palmprint segmentation approaches based on the online palmprints. One is designed to extract a fixed size square area as the central part sub-image for the feature extraction, while the remaining one using an inscribed circle. After that, some analysis has been performed to test the robustness of these approaches. It is shown that the circle-based approach can extract a larger area for the feature extraction. This is because the circle-based approach utilizes a large circle which is close to the boundary of a palm, while square-based approach only uses a fixed size 128×128 area. Also, by some experimental testing, the square-based approach cannot deal with some special palmprint samples whereas circle-based approach can handle it. In addition, the circle-based approach provides a means of palmprints classification which is useful in the large database. On the other hand, although the circle-based approach provides a larger surface area for the palmprint feature extraction, it may take longer time for feature extraction. We can say that square-based approach has the advantage of faster processing on the feature extraction while circle-based approach has more features obtained. We can conclude that the selection of a palmprint segmentation approach is dependent on the feature extraction algorithm and the application of the system. The circle-based approach is not stable since it is depending on palm size. Therefore, we take the square-based segmentation approach in the following chapters.

7 TEXTURE FEATURE USING 2-D GABOR FILTERS

This chapter reports a novel textured-based feature extraction method that uses low resolution online palmprint images for personal authentication. A palmprint is considered as a texture image, so an adjusted Gabor filter is employed to capture the texture information from palmprints. Section 7.2 describes the matching strategies of the proposed method. Section 7.3 discusses the parameter selection and rotational test. Section 7.4 reports the experimental testing and analysis. Section 7.5 summarizes this chapter.

7.1 Texture Feature Extraction

As we know, there are a few principal lines on a palmprint (see Section 1.4 of Chapter 1). Some algorithms such as the stack filter [37] can obtain these lines. However, the principal lines do not contribute adequately to high accuracy because of their similarity amongst different palms. So, wrinkles play an important role in the palmprint authentication. In fact, accurately extracting wrinkles is still a difficult task. This motivates us to apply texture analysis to palmprint authentication. Gabor filter, Gabor filter bank, Gabor transform and Gabor wavelet are widely applied to image processing, computer vision and pattern recognition. This function can provide accurate time-frequency location governed by the "Uncertainty Principle" [51-52]. A circular 2-D Gabor filter in the spatial domain has the following general form [53-54],

$$G(x, y, \theta, u, \sigma) = \frac{1}{2\pi\sigma^2} \exp\left\{-\frac{x^2 + y^2}{2\sigma^2}\right\} \exp\{2\pi i(ux\cos\theta + uy\sin\theta)\}, \quad (7\text{-}1)$$

where $i = \sqrt{-1}$; u is the frequency of the sinusoidal wave; θ controls the orientation of the function and σ is the standard deviation of the Gaussian envelope. Such Gabor filters have been widely used in various applications [55-64]. In addition to accurate time-frequency location, they also provide robustness against varying brightness and contrast of images. Furthermore, the filters can model the receptive fields of a simple cell in the primary visual cortex. Based on these properties, we can apply a Gabor filter to palmprint authentication in this chapter.

In fact, a Gabor function, $G(x, y, \theta, u, \sigma)$ with a special set of parameters (σ, θ, u), is transformed into a discrete Gabor filter, $G[x, y, \theta, u, \sigma]$. In order to provide more robustness to brightness, the Gabor filter is turned to zero DC (direct current) with the application of the following formula:

$$\tilde{G}[x,y,\theta,u,\sigma] = G[x,y,\theta,u,\sigma] - \frac{\sum_{i=-n}^{n}\sum_{j=-n}^{n}G[i,j,\theta,u,\sigma]}{(2n+1)^2}, \qquad (7\text{-}2)$$

where $(2n+1)^2$ is the size of the filter. Actually, the imaginary part of the Gabor filter automatically has zero DC because of odd symmetry. This texture feature extraction method has been applied to iris recognition [58].

The adjusted Gabor filter will convoluté with the preprocessed central part palmprint sub-image. This preprocessing approach is described in Chapter 6 Section 6.3 for palmprints alignment [50]. The central part palmprint sub-image, 128×128, is obtained to represent the whole palmprint. This palmprint segmentation technique reduces the translation and rotation of the palmprints captured from the same palms.

Palmprint Feature Representation
The sample point in the filtered image is coded to two bits, $(b_r,\ b_i)$, by the following inequalities:

$$b_r=1 \ \text{if} \ \ \text{Re}[\tilde{G}[x,y,\theta,u,\sigma]*I] \geq \ 0, \qquad (7\text{-}3)$$

$$b_r=0 \ \text{if} \ \ \text{Re}[\tilde{G}[x,y,\theta,u,\sigma]*I] < \ 0, \qquad (7\text{-}4)$$

$$b_i=1 \ \text{if} \ \ \text{Im}[\tilde{G}[x,y,\theta,u,\sigma]*I] \geq \ 0, \qquad (7\text{-}5)$$

$$b_i=0 \ \text{if} \ \ \text{Im}[\tilde{G}[x,y,\theta,u,\sigma]*I] < \ 0, \qquad (7\text{-}6)$$

where I is the central part palmprint sub-image. Using this coding method, only the phase information in palmprint images is stored in the feature vector.

7.2   Feature Matching

In order to describe clearly the matching process, each feature vector is considered as two 2-D feature matrices, real and imaginary. Palmprint matching is based on a normalized hamming distance. Let P and Q be two palmprint feature matrices. The normalized hamming distance can be defined as,

$$D_o = \frac{\sum_{i=1}^{N}\sum_{j=1}^{N}\left(P_R(i,j)\otimes Q_R(i,j)+P_I(i,j)\otimes Q_I(i,j)\right)}{2N^2}, \qquad (7\text{-}7)$$

where P_R (Q_R) and P_I (Q_I) are the real part and the imaginary part of P (Q), respectively; the Boolean operator, " $\otimes$ ", is equal to zero if and only if the two bits, $P_{R(I)}(i,j)$ and $Q_{R(I)}(i,j)$ are equal and the size of the feature matrices is $N{\times}N$. It is noted that D_o is between 1 and 0. The hamming distance for perfect matching is zero. In order to provide translation invariance matching, Equation 7-7 can be improved as

$$D_{min} = \min_{|s|<S,|t|<T} \frac{\sum_{i=\max(1,1+s)}^{\min(N,N+s)} \sum_{j=\max(1,1+t)}^{\min(N,N+t)} \left(P_R(i+s,j+t) \otimes Q_R(i,j) + P_I(i+s,j+t) \otimes Q_I(i,j) \right)}{2H(s)H(t)},$$

(7-8)

where $S=2$ and $T=2$ to control the range of horizontal and vertical translation of a feature in the matching process, respectively, and

$$H(s) = \min(N, N+s) - \max(1, 1+s).$$ (7-9)

The hamming distance, D_{min}, can support translation matching; nevertheless, because of unstable preprocessing, it is not a rotational invariant matching. Therefore, in enrollment mode, the coordinate system is rotated by a few degrees and then the sub-images are extracted for feature extraction. Finally, combining the effect of preprocessing and rotated features, Equation 7-8 can provide both approximately rotational and translation invariance matching.

7.3 Parameter Selection and Performance Test

The above sections give the architecture of the proposed method. In this section, parameter selection and rotational robustness are two main issues. Twelve filters with different parameters are selected to test their robustness. Gabor filters with these sets of parameters constitute a filter band and their real parts could be applied to texture analysis [55], listed in Table 7-1 for experimental testing.

Table 7-1. The parameters of the 12 filters.

Levels	No	Sizes	θ	u	σ
1	1	9 by 9	0	0.3666	1.4045
	2	9 by 9	45	0.3666	1.4045
	3	9 by 9	90	0.3666	1.4045
	4	9 by 9	135	0.3666	1.4045
2	5	17 by 17	0	0.1833	2.8090
	6	17 by 17	45	0.1833	2.8090
	7	17 by 17	90	0.1833	2.8090
	8	17 by 17	135	0.1833	2.8090
3	9	35 by 35	0	0.0916	5.6179
	10	35 by 35	45	0.0916	5.6179
	11	35 by 35	90	0.0916	5.6179
	12	35 by 35	135	0.0916	5.6179

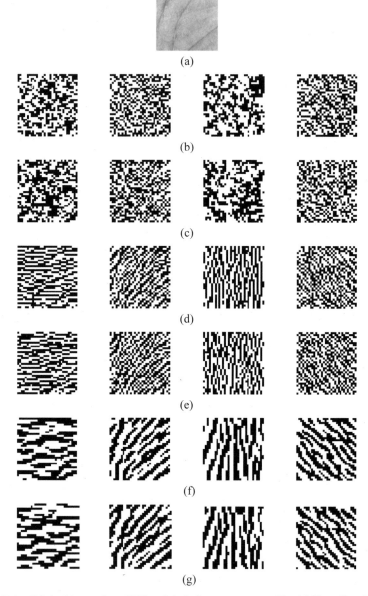

(a)

(b)

(c)

(d)

(e)

(f)

(g)

Figure 7-1. Original image from DBI and their features generated by 12 filters listed in Table
7-1. a) Original image, b), d) and f) real parts of features from Level 1, 2, and 3 filters,
respectively, c), e) and g) imaginary parts of from Level 1, 2 and 3 filters, respectively.

It should be pointed out that the success of 2-D Gabor phase coding depends on the
selection of Gabor filter parameters, θ, σ, and u. Also, we included an automatic
thresholding procedure to handle the misplacement of the palm during data sampling.

For presentation convenience, Filters 1-4 are marked as Level 1; Filters 5-8 and 9-12 are noted as Level 2 and Level 3, respectively. In order to investigate the relationship between feature size and accurate rate, two sets of images will be tested. Here we tried to use a small testing set of database for the parameter selection. But at the section of experimental testing, we use a larger database for the performance evaluation. The size of the first set is defined as 128 by 128 (called as DBI) and the size of the second set is defined as 64 by 64 (called as DBII), which is resized from the first one. The parameters, S and T, are 12 for DBI and 6 for DBII. The databases for testing contain 425 images from 95 persons, from **PolyU-ONLINE-Palmprint-II**. Figures 7-1 show the features generated by the given 12 filters.

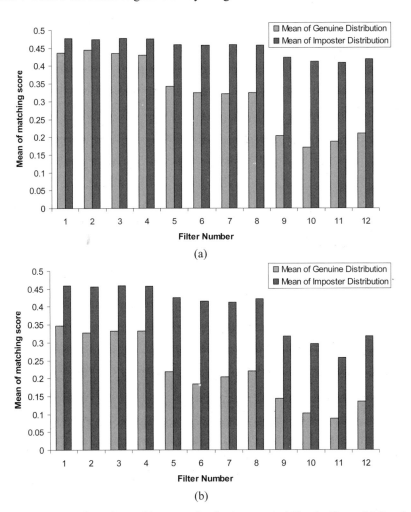

Figure 7-2. Mean of genuine and imposter distribution generated by the filters. (a) Results from DBI, and (b) Results from DBII

Accuracy Test

We apply the twelve filters to the two databases (DBI and DBII) to observe the relationship between parameters and accuracy. When σ increases and u decreases, we can obtain several properties from Figure 7-2, which shows the means of the imposter and genuine distributions on the two databases. The standard deviations of both imposter and genuine distributions on DBI and DBII are given in Figure 7-3. It is noted that the means of the imposter and genuine distributions decrease when σ increases and u decreases. As a result, the similarity between two palmprints increases no matter the two palmprints come from the same subject or not. According to Figure 7-3, it is clear that the standard deviation of the imposter distributions becomes larger. Intuitively, the number of the degree-of-freedom reduces because of the large size of the Gaussian envelope in the Gabor filter. The recognition rates of the test are shown in Table 7-2; the recognition rates using the Level 3 filters on DBI are similar to using the Level 2 filters on DBII. This is because the ratios of the length (width) of the images in DBI to σ in Level 3 filters and those in DBII to σ in Level 2 filters are the same. For DBI, in Table 7-2, the false rejection rate decreases when the size of filter and σ increase. However, this statement is not true for DBII. In the optimal case, using Filter 6 on DBII, the false rejection rate is 2.5% with 0% false acceptance rate. Different filters produce different sizes of feature, as mentioned in Table 7-3. The sizes of features are between 210 and 3,540 bytes.

Table 7-2. Summary of accuracy rate for the different filters on two databases.

Filter No.	DBI		DBII	
	False acceptance rate (%)	False rejection rate (%)	False acceptance rate (%)	False rejection rate (%)
1	0	41.4	0	17.9
2	0	41.1	0	15.0
3	0	68.8	0	12.4
4	0	42.4	0	15.8
5	0	17.7	0	5.0
6	0	19.2	0	2.5
7	0	13.9	0	7.2
8	0	46.4	0	5.9
9	0	5.6	0	28.2
10	0	2.9	0	18.7
11	0	7.3	0	36.0
12	0	6.4	0	29.3

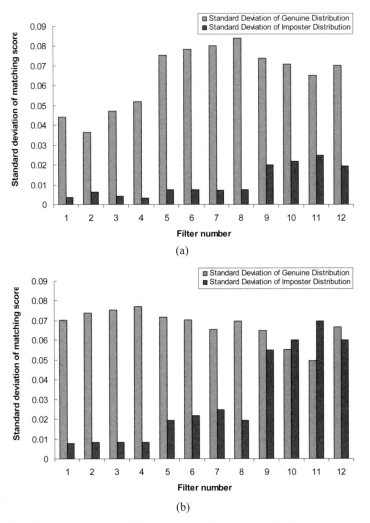

Figure 7-3. Standard deviation of the genuine and imposter distributions generated by the filters. (a) Results from DBI, and (b) results from DBII.

Table 7-3. Size of the defined features for the different filters and the databases.

Filter Level	Feature size in bytes	
	DBI	DBII
1	3540	756
2	3080	552
3	2162	210

Rotational Test

The rotational robustness of our method is investigated in this test. In each database, fifty images with different rotation degrees are selected and each of them will match with its original image. Same as the previous experiments, the twelve filters will be applied to the original images and the rotated images. The means of matching scores are illustrated in Figure 7-4. Obviously, the means of matching scores increase if the degree of rotation increases. Besides, according to the figure, a small image is more robust to rotation than a large one using the same filter. Level 3 filters are very robust to rotation for DBII; nevertheless, their recognition rates are very low.

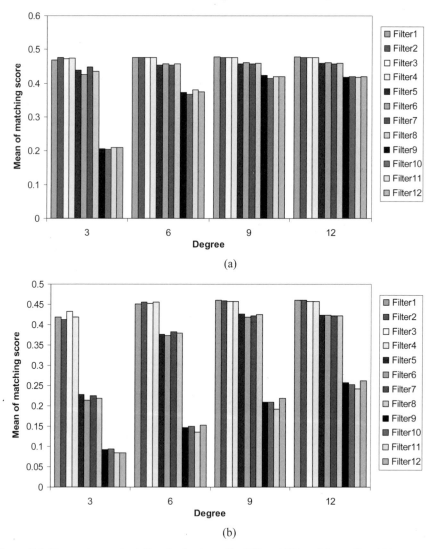

(a)

(b)

Figure 7-4. Comparisons of rotational robustness for different filters. Mean of matching scores from DBI and (b) Mean of matching scores from DBII.

7.4   Experimental Results and Analysis

A palmprint database of 4,647 palmprint samples (collected from 120 individuals) is formed from the **PolyU-ONLINE-Palmprint-II**, for palmprint experiments. 43 of them are female, 111 of them are less than 30 years old and 2 of them are more than 50 years old. Each of them is asked to provide about 10 images for their left palm and 10 images for their right palm in each of two occasions. In total, each subject provides about 40 images. The average time difference of the first and second occasions is 81 days. The maximum and minimum are 4 and 162 days, respectively. Originally, the collected images have two sizes, 384×284 and 768×568. The large images are resized to 384×284; consequently, the size of all the test images in the following experiments is 384×284 with 75dpi resolution. The central parts of each image extracted with size 128 by 128 are named DBI. The preprocessed images in DBI resized to 64 by 64 are named DBII. The DBII is used to test the possibility of using lower-resolution palmprint images for personal identification.

Verification Tests

To obtain best parameters for the proposed method, 12 different sets of parameters listed in Table 7-1 are used to test it again on this large database. Each of the images in DBI (DBII) is matched with all other palmprint images in the same database. A matching is counted as a correct matching if two palmprint images are collected from the same palm; otherwise it is an incorrect matching. The total number of matchings for one verification test is 10,794,981. 43,660 of them are correct matchings and rest of them are incorrect matchings. In total, 24 verification tests are carried out for testing the 12 sets of parameters on the two databases. The performance of different parameters on the two databases is presented by Receiver Operating Characteristic (ROC) curves, which are a plot of genuine acceptance rate against false acceptance rate for all possible operating points. Figures 7-5 (a)-(f) show the ROC curves for DBI (DBII) generated by Levels 1, 2, and 3 filters, respectively. According to the ROC curves, Level 3 filters are better than Levels 1 and 2 filters for DBI. According to Figure 7-5 (c), Filters 9, 10 and 11 provide similar performance when the false acceptance rate is greater than 10^{-2}. For false acceptance rates smaller than 10^{-2}, Filters 9 and 11 are better than Filter 10. For DBII, Level 1 filters are better than Level 2 and 3 filters. In fact, Filter 2 is the best for DBII. Although using very low resolution images as DBII's images cannot provide very good performance, it still gives us an insight into using very low resolution palmprint images for personal identification.

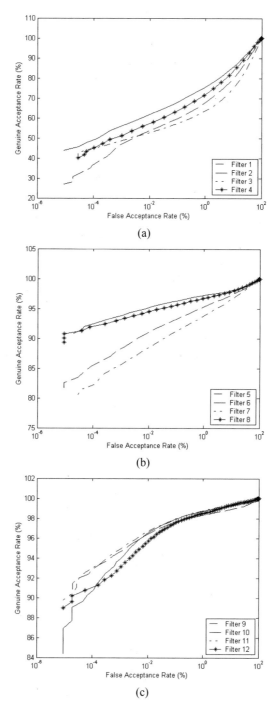

Figure 7-5. Verification test results. (a), (b) and (c), ((d), (e) and (f)) the ROC curves of Level 1, Level 2 and Level 3 filters from DBI (DBII), respectively.

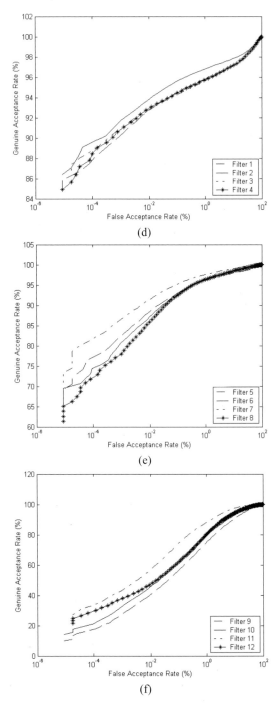

Figure 7-5. (Continued) Verification test results. (a), (b) and (c), ((d), (e) and (f)) the ROC curves of Level 1, Level 2 and Level 3 filters from DBI (DBII), respectively.

Speed

The algorithm was implemented in Visual C++ 6.0 on a PC using Intel Pentium III processor (500MHz). We obtain the execution time for the preprocessing, feature extraction and matching, as shown in Table 7-4. The total execution time is about 0.6 second, which is fast enough for real-time verification. For identification, if the database contains 100 persons and each person registers 3 palmprint images, the total identification time is about 1.1s. In fact, we have not optimized the code so that further reduction in the computation time is possible.

Table 7-4. Execution time for palmprint verification system.

Operation	Execution Time
Preprocessing	538ms
Feature Extraction	64ms
Matching	1.7ms

Comparisons with Other Biometrics

We first compare our feature extraction method with offline and online palmprint approaches [18, 31]. Then, similar comparisons with iris and fingerprint [59, 117] are discussed. Offline palmprint identification uses lines as feature [18, 77, 82], which requires a set of filters or operations to extract line and point information. The offline images are usually of high resolution. Our proposed method only uses two filters to extract texture information. Based on the above two reasons (lower resolution and less filters used), feature extraction using our proposed method is faster than the offline approach. The discrimination power of our texture-based method is also very high because it can handle palmprint images with unclear lines (see Figure 7-6) and similar principal lines (see Figure 7-7). Importantly, the accuracy rate of the method is higher than that of the offline one and our method can support real-time applications.

Iris and fingerprint using similar method for the feature extraction, are marked as IrisCode and FingerCode, respectively. Table 7-5 summarizes their performance. In fact, our proposed method (PalmCode) only requires very small image so the computation time for filtering is very short. In terms of accuracy, IrisCode is the best. However, it suffers from high costs of input devices. According to the current experimental results, PalmCode is better than FingerCode.

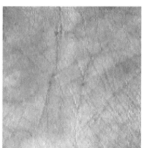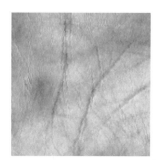

Figure 7-6. Three typical images with unclear lines.

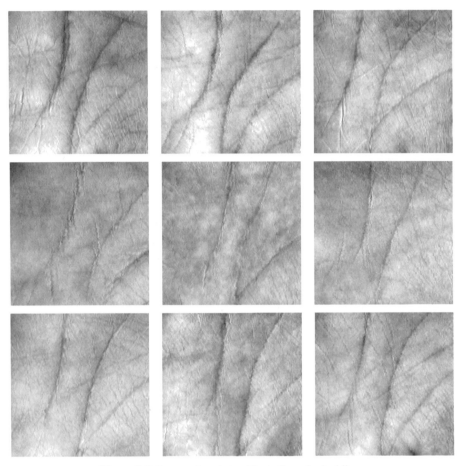

Figure 7-7. Some palmprints with similar principal lines.

7.5 Summary

This chapter reports a novel textured-based feature extraction method using low resolution online palmprint images for personal authentication. A palmprint is considered as a texture image, so an adjusted Gabor filter is employed to capture the texture information on palmprints. Based on our tests, some filters (such as Filter 11) can be selected in our palmprint feature extraction, which is robust in terms of accuracy and other consideration. Combined with the effects of preprocessing and rotated preprocessed images, our matching process is translation and rotational invariance. Experimental results illustrate the effectiveness of the method.

Table 7-5. Comparisons of PalmCode, IrisCode and FingerCode.

	PalmCode	IrisCode [116]	FingerCode [59]
Feature size	552 bytes	256 bytes or 512 bytes [#]	640 bytes or 896 bytes
Number of filter using for feature extraction	2 (real and imaginary)	Multiple	8
Filter size	17 by 17	Unknown	33 by 33
Feature extracted from filtered image	Phase	Phase	Average absolute deviation
Translation invariance	Yes	Yes	Yes
Rotational invariance	Yes	Yes	Approximate
Cost of input equipment	Low	High	Low
User acceptance	High	Depending on capture device	High

Originally, the size of IrisCode is 256. The current products in the market use extra 256 bytes to mark the eyelashes, reflections and boundary artifacts of hand contact lenses.

8 LINE FEATURES EXTRACTION AND REPRESENTATION

Different researchers have put their efforts into palm line extraction. Han [66] used Sobel and morphological operations to extract line-like features from palmprints images obtained using a scanner. Similarly, Kumar [67] used directional masks to extract line-like features from palmprints captured using a digital camera. Neither Han nor Kumar explicitly extracted palm lines from the online palmprints. None of these online palmprint recognition methods used palm line structural features explicitly, yet palm line structural features can describe a palmprint clearly and are robust against illumination and noise. In this chapter, we report a set of directional line detectors based on the properties of palm lines. To avoid losing the details of the palm line structure, these irregular lines are represented by their chain code. Section 8.2 reports palm line matching strategies which compute a matching score between two palmprints according to the points of their palm lines. Section 8.3 discusses experimental results and analysis. Section 8.4 summarizes this chapter.

8.1 Palm Line Feature Extraction

Palm lines, including the principal lines and wrinkles, are a kind of roof edge. A roof edge is generally defined as a discontinuity in the first-order derivative of a gray-level profile [68]. In other words, the positions of roof edge points are the zero-cross points of their first-order derivative. The magnitude of the edge points' second-derivative can reflect the strength of these edge points [69]. These properties can be used to detect palm lines. The directions of palm lines are arbitrary and it is very difficult to obtain them directly from noisy images. In this section, we detect palm lines in different directions. The lines that are detected in θ direction are called *directional lines in θ direction*.

Suppose that $I(x, y)$ denotes an image. We devise the horizontal line detector (the directional line detector in $0°$ direction). To improve the connection and smoothness of the lines, the image is smoothed along the line direction (horizontal direction) using a 1-D Gaussian function G_{σ_s} with variance σ_s:

$$I_s = I * G_{\sigma_s} \qquad (8\text{-}1)$$

where * is the convolve operation.

The first- and the second-order derivatives in the vertical direction can be computed by convolving the smoothed image with the first- (G'_{σ_d}) and second- (G''_{σ_d}) order derivative of a 1-D Gaussian function G_{σ_d} with variance σ_d:

$$I' = I_s * (G'_{\sigma_d})^T = (I * G_{\sigma_s}) * (G'_{\sigma_d})^T = I * (G_{\sigma_s} * (G'_{\sigma_d})^T) = I * H_1^0, \quad (8\text{-}2)$$

$$I'' = I_s * (G''_{\sigma_d})^T = (I * G_{\sigma_s}) * (G''_{\sigma_d})^T = I * (G_{\sigma_s} * (G''_{\sigma_d})^T) = I * H_2^0, \quad (8\text{-}3)$$

where $H_1^0 = G_{\sigma_s} * (G'_{\sigma_d})^T$, $H_2^0 = G_{\sigma_s} * (G''_{\sigma_d})^T$; T is the transpose operation ; $*$ is the convolve operation. H_1^0, H_2^0 are called the horizontal line detectors (*directional line detectors in 0^0 direction*). The horizontal lines can be obtained by looking for the zero-cross points of I' in the vertical direction and their strengths are the values of the corresponding points in I'':

$$L_0^1(x,y) = \begin{cases} I''(x,y), & \text{if } I'(x,y) = 0 \text{ or } I'(x,y) \times I'(x+1,y) < 0; \\ 0 & \text{otherwise.} \end{cases} \quad (8\text{-}4)$$

Furthermore, we can determine the type of a roof edge (line), i.e. valley or peak, from the sign of the values in $L_0^1(x,y)$: plus signs (+) represent valleys, while minus signs (-) represent peaks. Since all palm lines are valleys, the minus values in $L_0^1(x,y)$ should be discarded:

$$L_0^2(x,y) = \begin{cases} L_0^1(x,y), & \text{if } L_0^1(x,y) > 0; \\ 0, & \text{otherwise.} \end{cases} \quad (8\text{-}5)$$

Palm lines are much thicker than ridges. For that reason, one or more thresholds can be used to remove ridges from L_0^2 to obtain a binary image L_0, which is called the directional line image in 0^0 direction. The directional line detectors H_1^θ, H_2^θ in θ direction can be obtained by rotating H_1^0, H_2^0 with angle θ. The line points can be obtained by looking for the zero-cross points in $\theta + 90^\circ$ direction. After discarding the peak roof edges and thresholding, we can obtain the binary directional line image L_θ in θ direction. Finally, all of the directional line images are combined to obtain a line image, denoted as L , as below:

$$L(i,j) = \bigvee_{\text{all } \theta} L_\theta(i,j), \quad (8\text{-}6)$$

where $\vee$ is a logical "OR" operation. After conducting the closing and thinning operations, we obtain the resultant palm line image. Figure 8-1 shows the process of palm line extraction: (a) is an original palmprint image; (b)-(e) indicate directional line images in 0°, 45°, 90° and 135° directions, respectively; (f) is the resultant palm line image and (g) shows the original palmprint overlapped with the extracted palm lines.

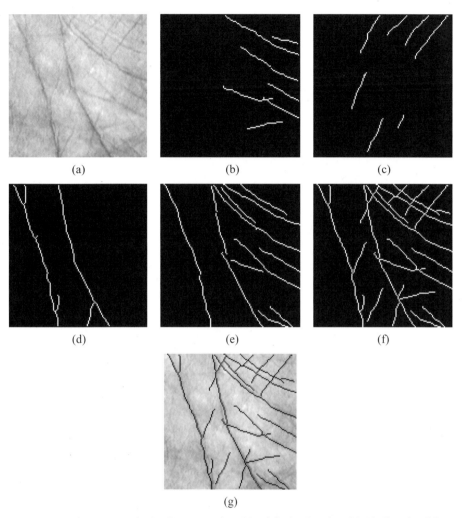

Figure 8-1. The process of palm line extraction: (a) original palmprint, (b)-(e) directional line images in 0°, 45°, 90° and 135° ($\sigma_s = 1.8, \sigma_d = 0.5$), (f) resultant palm line image, and (g) original palmprint overlapping with the extracted palm lines.

There are two parameters σ_s and σ_d (see Equations 8-1 and 8-2) in these directional line detectors. σ_s controls the connection and smoothness of the lines and σ_d handles the width of the lines which can be detected. Small σ_s results in poor connection and poor smoothness of the detected lines while large σ_s results in the loss of some short lines and of line segments whose curvatures are large. Thin roof edges cannot be extracted when σ_d is large. So we should choose the appropriate parameters *ad hoc*. In general, palm lines are long, narrow and somewhat straight, thus σ_s should be large while σ_d should be small for palm line extraction.

After tuning the values of σ_s and σ_d through a lot of experiments, we obtained the suitable ones for palm line extraction: $\sigma_s = 1.8$ and $\sigma_d = 0.5$. In this case,

$$H_1^0 = \begin{bmatrix} 0.0009 & 0.0027 & 0.0058 & 0.0092 & 0.0107 & 0.0092 & 0.0058 & 0.0027 & 0.0009 \\ 0.0065 & 0.0191 & 0.0412 & 0.0655 & 0.0764 & 0.0655 & 0.0412 & 0.0191 & 0.0065 \\ 0.0000 & 0.0000 & 0.0000 & 0.0000 & 0.0000 & 0.0000 & 0.0000 & 0.0000 & 0.0000 \\ -0.0065 & -0.0191 & -0.0412 & -0.0655 & -0.0764 & -0.0655 & -0.0412 & -0.0191 & -0.0065 \\ -0.0009 & -0.0027 & -0.0058 & -0.0092 & -0.0107 & -0.0092 & -0.0058 & -0.0027 & -0.0009 \end{bmatrix} \tag{8-7}$$

$$H_2^0 = \begin{bmatrix} 0.0156 & 0.0211 & 0.0309 & 0.0416 & 0.0464 & 0.0416 & 0.0309 & 0.0211 & 0.0156 \\ 0.0257 & 0.0510 & 0.0954 & 0.1441 & 0.1660 & 0.1441 & 0.0954 & 0.0510 & 0.0257 \\ -0.0298 & -0.1125 & -0.2582 & -0.4178 & -0.4896 & -0.4178 & -0.2582 & -0.1125 & -0.0298 \\ 0.0257 & 0.0510 & 0.0954 & 0.1441 & 0.1660 & 0.1441 & 0.0954 & 0.0510 & 0.0257 \\ 0.0156 & 0.0211 & 0.0309 & 0.0416 & 0.0464 & 0.0416 & 0.0309 & 0.0211 & 0.0156 \end{bmatrix} \tag{8-8}$$

Figure 8-2 shows some examples of the extracted palm lines in which (a) is the original palmprint images; (b) is the extracted palm lines and (c) is the palmprint images overlapped with the extracted palm lines. In these examples, four directional line detectors (0°, 45°, 90° and 135°) are used for palm line extraction.

Because palm lines are formed in irregular patterns, it is very difficult to represent them in a mathematical form. An efficient method for representing irregular lines is chain code [70]. A direction code (0, 1, ..., 7) is defined for each point on a line according to its neighbor point and the chain code of the line is the direction code list of the points on the line. Each line can be represented by the coordinate of its beginning point and its chain code. This representation allows the palm lines to be easily and accurately restored.

8.2 Line Matching Strategies

Let L denotes a palm line image and C denotes a template represented by chain codes. The most natural way to match L against C is to count the line points that are at the same position in L and the palm line image, L_C, restored from C. However, because of the existence of noise, the line points of the same lines may be not superposed in the palmprints captured from the same palm in different time. To overcome this problem, we can first dilate L to get L_D and then count the overlapping points. The matching score between L and C is defined as below:

$$S(L,C) = \frac{2}{M_L + M_C} \times \sum_{i=1}^{M} \sum_{j=1}^{N} [L_D(i,j) \wedge L_C(i,j)], \tag{8-9}$$

where "$\wedge$" is the logical "AND" operation; $M \times N$ is the size of L; M_C and M_L are the number of non-zero points in L_C and L, respectively.

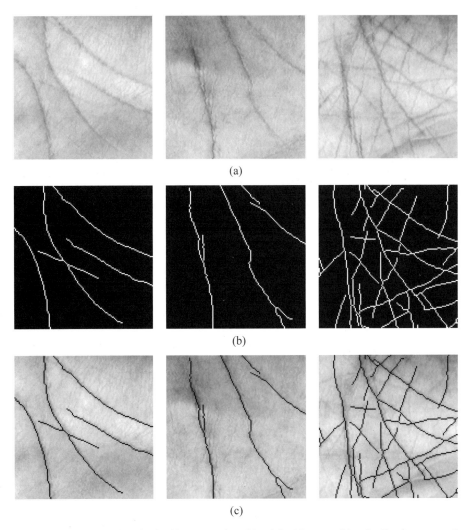

Figure 8-2. Some results of palm line extraction: (a) original images, (b) palm line images, and (c) original images overlapping with the extracted palm lines.

Table 8-1. Matching scores between palmprints in Figure 8-2.

	Left Column	Middle Column	Right Column
Left Column	1	0.1816	0.2046
Middle Column	0.1742	1	0.1902
Right Column	0.1840	0.1866	1

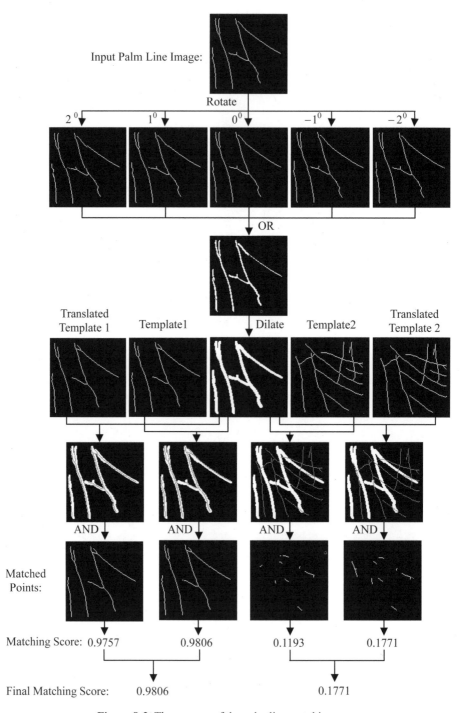

Figure 8-3. The process of the palm line matching.

From Equation 8-9, we have to scan the whole image to compute the matching score, which is time consuming. In fact, Equation 8-9 is equivalent to the following equation:

$$S(L,C) = \frac{2}{M_L + M_C} \times \sum_{i=1}^{M_C} L_D(x_i, y_i), \qquad (8\text{-}10)$$

where L_D is the dilated image of L; M_C is the total number of the line points restored from C and $(x_i, y_i)(i = 1, \cdots, M_C)$ are their coordinates; and M_L is the number of line points in L. Obviously, the computational complexity of Equation 8-10 is much less than that of Equation 8-9. Moreover, if we use Equation 8-10 to compute the matching score, instead of restoring the palm line image, L_C, from C, we only need to restore the coordinates of the line points. It should be noted that $S(L,C)$ is not symmetrical, that is $S(L_1,C_2)$ may not equal $S(L_2,C_1)$, where L_1, L_2 are two palm line images and C_1, C_2 are their chain codes, respectively. However, in general, the difference between $S(L_1,C_2)$ and $S(L_2,C_1)$ is small.

Obviously, $S(L,C)$ is between 0 and 1 and the larger the matching score, the greater the similarity between L and C. The matching score of a perfect match is 1. However, due to the imperfect preprocessing of palmprints, there may still be a little rotation and translation between the palmprints captured from the same palm at different times. To minimize the rotation effect, we rotate L with some degrees and then merge all the rotated images by using logical "OR" to get a merged binary image L_M. After that L_M is dilated to obtain L_{MD}. We will use L_{MD} to compute the matching score. To overcome the translation problem, we vertically and horizontally translate C by some points, and then at each translated position, compute the matching score between the translated C and L_{MD}. Finally, the maximum score among the matching scores of all translated positions is regarded as the final matching score. Figure 8-3 shows the entire process of palm line matching. In this figure, the input palm line image and Template 1 are from one palm while Template 2 is from another palm. The matching scores between the input sample and Template 1 and between the input sample and Template 2 are 0.9806 and 0.1771, respectively. The matching scores between the palmprints in Figure 8-2 are listed in Table 8-1.

8.3   Experiments and Analysis

When palmprints are captured, the position, direction and amount of stretching of a palm may vary so that even palmprints from the same palm may have a little rotation and translation. Furthermore, palms differ in size. Hence palmprint images should be orientated and normalized before feature extraction and matching. In our CCD based palmprint capture device [65], there are some pegs between fingers to guides the palm's stretching, translation and rotation. These pegs separate the fingers, forming holes between the index finger and the middle finger, and between the ring finger and the little finger. In this chapter, we use the preprocessing technique described in

Chapter 6 Section 6.3 to align the palmprints [50]. In this technique, the tangent of two finger holes are computed and used to align the palmprint. The central part palmprint sub-image, 128 × 128, is obtained to represent the whole palmprint. Such preprocessing greatly reduces the translation and rotation of the palmprints captured from the same palms.

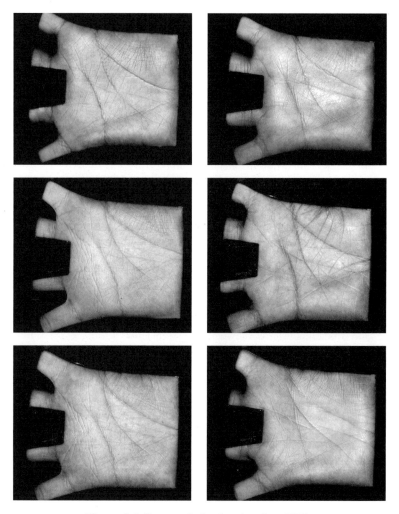

Figure 8-4. Some typical palmprints from DB1.

Three palmprint databases, DB1, DB2 and DB3, were established for our experiments. A large palmprint database, DB1, is formed by taking 7,605 palmprint samples from the **PolyU-ONLINE-Palmprint-II**, for the palmprint matching and verification tests. There are two different sizes of palmprints from this database, 384 × 284 and 768 × 568. Figure 8-4 shows some samples from DB1. We also collected two smaller databases, DB2 and DB3, which contain 400 palmprints from 40 different palms, respectively. These two smaller databases were established as follows: 1) 40 palms were asked to provide 400 images (10 images per palm) to build DB2 when they were clean, and 2) these palms were made dirty and provided other 400 images (also 10 images per palm) to build DB3.

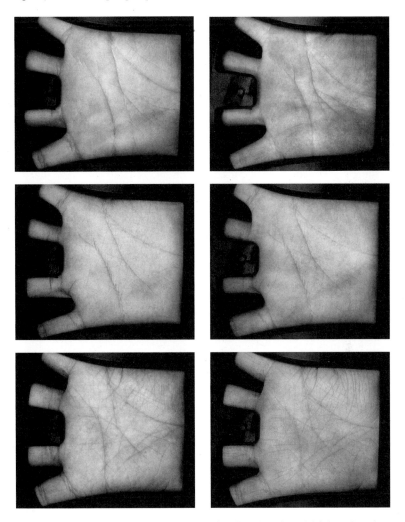

Figure 8-5. Some samples from DB2 (left column) and DB3 (right column).

These two databases were used to investigate the dirty palms' effect upon the performance of the proposed approach. All of the images in both DB2 and DB3 are 768 × 568. Figure 8-5 shows some palmprints from DB2 and DB3, respectively. In our experiments, all images were resized to 384 × 284, and using the preprocessing technique proposed described in Chapter 6 Section 6.3, in which the central part palmprint sub-image, 128 × 128, is obtained to represent the whole palmprint.

Four directional line detectors (0°, 45°, 90° and 135°), which were computed according to H_1^0 and H_2^0 in Equations 8-7 and 8-8, were used to extract palm lines. In addition, the Hysteresis threshold method [71] was used. In this method, the high threshold is chosen automatically by using Otsu's method [72] to the non-zero points of all L_θ^1 and the low threshold was chosen as the minimum value of the non-zero points of all L_θ^1. In the palm line matching process, the ranges of palmprint rotation were -2^0 to 2^0 and the ranges of vertical and horizontal translation were −5 to 5 pixels.

Palmprint Matching

To test the performance of the proposed approach, each sample in DB1 is matched against the other palmprints in the same database. The matching between palmprints which were captured from the same palm is defined as a correct matching. Otherwise, the matching is defined as an incorrect matching. A total of 57,828,420 (7,605 × 7,604) matchings have been performed, in which 142,686 matchings are correct matchings.

Figure 8-6 shows the distributions of correct and incorrect matching scores. It is shown that there are two distinct peaks in the distributions of the matching scores. One peak (located at around 0.80) corresponds to correct matching scores while the other peak (located at around 0.23) corresponds to incorrect matching scores. These two peaks are widely separated and the distribution curve of the correct matching scores intersects very little with that of incorrect matching scores. The proposed approach effectively discriminate palmprints from different persons.

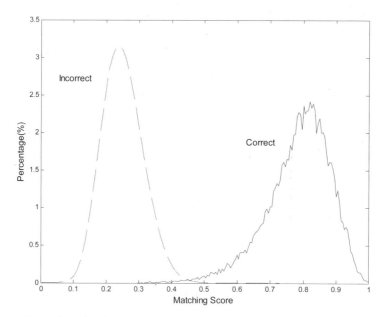

Figure 8-6. The distributions of the correct and incorrect matching scores.

Palmprint Verification

Palmprint verification, also called one-to-one matching, involves answering the question "Whether this person is whom he or she claims to be" by examining his or her palmprint. In palmprint verification, a user indicates his or her identity, thus the input palmprint is only matched against his or her stored template. To determine the accuracy of the verification, each sample in DB1 is matched against the other palmprints in the same database. If the matching score of the sample palmprint exceeds a given threshold, it is accepted. If not, it is rejected. The performance of a verification method is often measured by the false acceptance rate (FAR) and false rejection rate (FRR). While it is ideal that these two rates should be as low as possible, they cannot be lowered at the same time. So, depending on the application, it is necessary to make a tradeoff: for high security systems, such as some military systems, where security is the primary criterion, we should reduce FAR, while for low security systems, such as some civil systems, where ease-of-use is also important, we should reduce FRR. To test the performance of a verification method with respect to the FAR/FRR tradeoff, we usually plot the so-called *Receiver Operating Characteristic* (ROC) curve, which plots the FAR against the FRR. Figure 8-7 shows the ROC curves of the proposed approach and of the 2-D Gabor algorithm [65], which was implemented in DB1. The proposed approach's equal error rate (EER) is about 0.4% (i.e. FAR equals to FRR), while the EER of the 2-D Gabor algorithm is about 0.6%. From this figure, the proposed approach has a smaller FAR except when FRR exceeds 0.91%. This means that the 2-D Gabor algorithm is more suitable for high security systems while the proposed approach is more suitable for medium and low security system requirements.

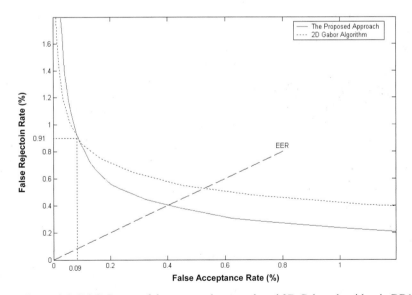

Figure 8-7. ROC Curves of the proposed approach and 2D Gabor algorithm in DB1.

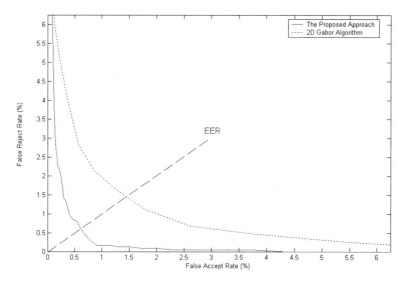

Figure 8-8. ROC Curves of the proposed approach and 2D Gabor algorithm when the dirty palmprints match against the clean palmprints.

Dirty Palms: Their Effect upon Performance

To investigate how dirty palms might affect the performance of the proposed approach, we matched each image in DB3, captured from dirty palms, against each image in DB2, captured from clean palms. Figure 8-8 shows the ROC curves of the proposed approach and 2-D Gabor algorithm [65], where their EERs are 0.61% and 1.24%, respectively. That is, the EER of our approach is less 0.63% than that of 2-D Gabor algorithm. This figure shows that the whole ROC curve of the proposed approach is below the curve of the 2-D Gabor algorithm, which means that the proposed approach outperforms the 2-D Gabor algorithm. In other words, our approach is more robust to dirty palms than 2-D Gabor algorithm. This is because texture features as extracted in the 2-D Gabor algorithm are more easily obscured by dirt than are line features of the type extracted using the proposed approach.

Storage Requirements

Storage requirements are very important for biometric systems. The proposed approach represents palmprints using their palm line chain codes. Each palm line is stored with the coordinates of one of its endpoints and its chain code. Suppose the length of a palm line is L points, the length of its chain code will be $(L-1)$. Since 3 bits are needed to store a direction code $(0, 1, ..., 7)$, $3 \times (L-1)$ bits are needed to represent the whole chain code. The coordinates of one endpoint of the palm line have two components and one byte is needed to store each component. Therefore, a total of $[2 + 3 \times (L-1)/8]$ bytes are needed to store each palm line. The statistical results from DB1 show that the average number of the palm lines of a palmprint is about 22 and the average length of a palm line is about 28 points. Consequently, the average storage requirement for a palmprint is about $22 \times [2 + 3 \times (28 - 1)/8] \approx 267$

bytes. The maximum and minimum storage requirements in these palmprints are 762 bytes and 62 bytes, respectively. The storage requirement of the proposed approach is listed in Table 8-2.

Table 8-2. Storage requirements of the proposed approach.

	Bytes needed
Maximum	762
Minimum	62
Average	267

Speed

Speed is another key factor affecting the selection of a biometric system. The proposed approach is implemented using Visual C++ on a personal computer with Intel Pentium III processor (500MHz). The average time that is required to preprocess, extract and match palmprint is listed in Table 8-3. The average response time is about 0.6 second for verification, which is fast enough for a real time biometric system.

Table 8-3. Average time taken using the proposed approach.

	Time (ms)
Preprocessing	538
Palm Line Extraction	82
Palmprint Matching	2.2

8.4 Summary

Palm lines are one of the most important features of palmprints. In this chapter we have proposed a novel approach on palm line extraction and matching for personal authentication. A set of directional line detectors has been devised for effective palm line extraction. To preserve the details of the lines' structure, palm lines are represented by their chain code and then palmprints are matched by matching the points on their palm lines. The experiment results from a general database (DB1) demonstrate the proposed approach is more powerful for palmprint verification than the 2-D Gabor algorithm in [65] when FRR < 0.91%, and the EER of our approach is also decreased from 0.6% of 2-D Gabor algorithm to 0.4%. Another experiment for testing its robustness against dirty palms (DB3) confirms the advantage of our approach over 2-D Gabor algorithm, where the EER of our approach is 0.63% less than that of 2-D Gabor algorithm. In addition, the average memory requirement for a palmprint is only 267 bytes and the average processing time, including preprocessing of palmprint image and matching, is only 0.6 second, which has proved the practical value of our approach. In conclusion, the proposed approach can effectively identify a person based on his/her palm lines which can be used on a real world application.

9 LINEAR DISCRIMINATION FEATURE EXTRACTION

This chapter discusses the extraction of algebraic features from palmprint for use in identity recognition. Algebraic features, which represent intrinsic attributions of an image, can be obtained using various algebraic transformations or matrix decompositions [81]. Section 9.1 defines the algebraic feature known as a Fisherpalm [136]. An analysis for fisherpalms is performed in Section 9.2. Section 9.3 illustrates another algebraic feature, an eigenpalm, which is based on the eigenvectors of the covariance matrix. Section 9.4 reports the experimental results and analysis of the eigenpalms. Section 9.5 summarizes this chapter.

9.1 Definition: Fisherpalms

Fisher's linear discriminant (FLD) [76] is based on the linear projections to seek for the directions which are advantageous for discrimination. In other words, the class separation is maximized in these directions. Figure 9-1 illustrates intuitively the principal of FLD. In this figure, the samples, which are in the 2D feature space, are from two classes. Obviously, it is difficult to tell apart them in the original 2D space. However, if we use FLD to project these data from 2D to 1D, we can easily to discriminate them in such 1D space. This approach has been widely used in pattern recognition. Fisher [78] first proposed this approach for taxonomic classification. Cui et al. [75] employed a similar algorithm (they called it the Most Discriminating Feature – MDF) for hand sign recognition. Liu [81] adopted FLD to extract the algebraic features of handwritten characters. Belhumner et al. [73] developed a very efficient approach (Fisherfaces) for face recognition. This section presents a novel palmprint recognition method, called Fishpalms, based on FLD. In this method, each palmprint image is considered as a point in a high-dimensional image space. A linear projection based on FLD is used to project palmprints from this high-dimensional space to a significantly lower dimensional feature space, in which the palmprints from the different palms can be discriminated much more efficiently.

Fisherpalms Feature Extraction

A $N \times N$ palmprint image can be considered as a N^2 vector in which each pixel corresponds to a component. That is, $N \times N$ palmprint images can be regarded as points in a high-dimensional space (N^2–dimensional space), called the *original palmprint space (OPS)*. Generally, the dimension of the OPS is too high to be used directly. For example, the dimension of the original 128×128 palmprint image space is 16,384. Therefore, we should reduce the dimension of the palmprint image and, at the same time, improve or keep the discriminability between palmprint classes. A linear projection based on Fisher's Linear Discrimant (FLD) is selected for this purpose.

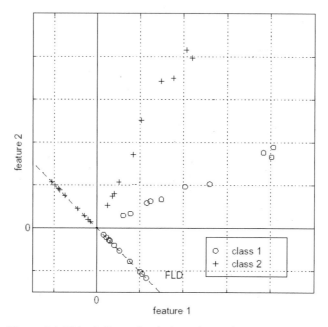

Figure 9-1. Fisher's linear discriminant (FLD) for a two-class problem.

Let us consider a set of N palmprints $\{x_1, x_2, \cdots, x_N\}$ taking values in a n-dimensional OPS, and assume that each image is captured from one of c palms $\{X_1, X_2, \cdots, X_c\}$ and the number of images from $X_i (i = 1, 2, \cdots, c)$ is N_i. FLD tries to find a linear transformation W_{opt} to maximize the Fisher criterion [76]:

$$J(W) = \frac{|W^T S_B W|}{|W^T S_W W|},\qquad (9\text{-}1)$$

where S_B and S_W are the between-class scatter matrix and the within-class scatter matrix:

$$S_B = \sum_{i=1}^{c} N_i (\mu_i - \mu)(\mu_i - \mu)^T,\qquad (9\text{-}2)$$

$$S_W = \sum_{i=1}^{c} \sum_{x_k \in X_i} (x_k - \mu_i)(x_k - \mu_i)^T,\qquad (9\text{-}3)$$

$$\mu_i = \frac{1}{N_i} \sum_{x_l \in X_i} x_l, \tag{9-4}$$

$$\mu = \frac{1}{N} \sum_{j=1}^{N} x_j. \tag{9-5}$$

The optical linear transformation W_{opt} can be obtained as follows [73, 76]:

$$W_{opt} = \arg\max_W \frac{|W^T S_B W|}{|W^T S_W W|} = [w_1, w_2, \cdots, w_m], \tag{9-6}$$

where $\{w_i \mid i = 1,2,\cdots,m\}$ is the set of generalized eigenvectors of S_B and S_W corresponding to the m nonzero generalized eigenvalues $\{\lambda_i \mid i = 1,2,\cdots,m\}$, i.e.

$$S_B w_i = \lambda_i S_W w_i, \qquad i = 1,2,\cdots,m. \tag{9-7}$$

There are at most c-1 nonzero generalized eigenvalues [76], hence the upper bound of m is $c-1$, where c is the number of palmprint classes.

Obviously, up to now, our discussion is based on the assumption that the denominator of Equation 9-1 does not equal to zero, i.e. S_W is of full rank. However, in general, S_W is not a full rank matrix. This stems from the fact that the rank of S_W is at most $N - c$, and, in general, the number of images in the training set N is much smaller than the number of pixels in each image n. This means that it is possible to choose the matrix W_{opt} such that the within-class scatter of the projected samples, i.e. the denominator of Equation 9-1, can be made exactly zero. Thus we cannot use Equations 9-1, 9-6 and 9-7 to obtain W_{opt} directly. In order to overcome this problem, the original palmprint images are firstly projected to a lower dimensional space by using K-L transformation so that the resulting within-class scatter matrix is nonsingular. And then the standard FLD is employed to process the projected samples. This method, which has been used efficiently in face recognition [73], is described as below:

1) Compute the transformation matrix of K-L transformation U_{KL}:

$$U_{KL} = \arg\max_U |U^T S_T U| = [u_1, u_2, \cdots, u_n], \tag{9-8}$$

where $\{u_i \mid i = 1,2,\cdots,n\}$ is the set of eigenvector of S_T corresponding to the nonzero eigenvalues and S_T is the total scatter matrix defined as:

$$S_T = \sum_{k=1}^{N} (x_k - \mu)(x_k - \mu)^T. \tag{9-9}$$

2) Compute the transformed within-class scatter matrix S'_W, which is a full rank matrix:

$$S'_W = U_{KL}^T S_W U_{KL}. \tag{9-10}$$

3) Compute the transformed between-class scatter matrix S'_B:

$$S'_B = U_{KL}^T S_B U_{KL}. \tag{9-11}$$

4) The standard FLD defined by Equation 9-6 is used to the transformed samples to obtain W_{fld}:

$$W_{fld} = \arg\max_{W} \frac{|W^T S'_B W|}{|W^T S'_W W|} = \arg\max_{W} \frac{|W^T U^T_{KL} S_B U_{KL} W|}{|W^T U^T_{KL} S_W U_{KL} W|}. \tag{9-12}$$

5) Compute W_{opt}:

$$W^T_{opt} = W^T_{fld} U^T_{KL}. \tag{9-13}$$

The columns of W_{opt} $\{w_1, w_2, \cdots, w_m\}$ $(m \leq c-1)$ are orthonormal vectors. The space spanned by these vectors is called the *Fisher palmprint space (FPS)* and the vector $w_i (i = 1,2,\cdots,m; m \leq c-1)$ is called a *Fisherpalm* [136]. Figure 9-2 shows an example of the Fisherpalm in the case of two palmprint classes. In this case, there is only one Fisherpalm in W_{OPT}. In this figure, (a) and (b) are samples in the class one and two, respectively, and (c) is the Fisherpalm used for classification.

The block diagram of the Fisherpalms based palmprint recognition is shown in Figure 9-3. There are two stages in our system: the enrollment stage and the recognition stage. In the enrollment stage, the Fisherpalms W_{opt} are computed by using the training samples using Equations 9-1 to 9-13, and stored as a FPS at first, and then the mean of each palmprint class is projected onto this FPS:

$$T = W^T_{opt} M, \tag{9-14}$$

where $M = \{m_1, m_2, \cdots, m_c\}$, c is the number of palmprint classes and each column of M, m_i $(i = 1,2,\cdots,c)$, is the mean of the ith class palmprints. T is stored as the template for each palmprint class. In the recognition stage, the input palmprint image is projected onto the stored FPS to get its feature vector V, and then V is compared with the stored templates to obtain the recognition result.

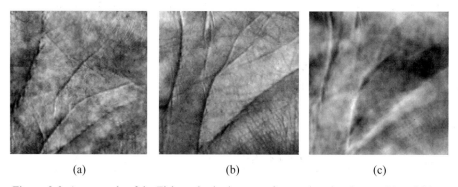

(a)   (b)   (c)

Figure 9-2. An example of the Fisherpalm in the case of two palmprint classes: (a) and (b) are samples in the class one and two, respectively, and (c) is the Fisherpalm used for classification.

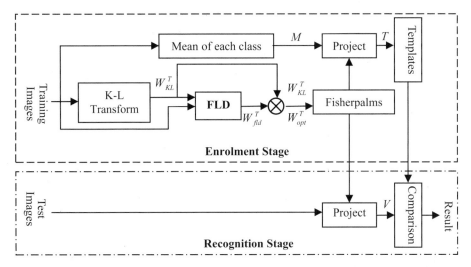

Figure 9-3. Block diagram of the Fisherpalms based palmprint recognition.

9.2 Analysis on Fisherpalms

A palmprint database of 3,000 palmprint samples is formed from the **PolyU-ONLINE-Palmprint-II**, for palmprint experiments testing. There are 300 different palms in which each palm is captured for 10 times in one occasion. The resolution of all of the original palmprint images is resized to 384×284 pixels, at 75 dpi. By using the preprocessing approach described in Chapter 6 Section 6.3, the tangents of two finger holes are computed for palmprints alignment. The central part palmprint sub-image, 128×128, is obtained to represent the whole palmprint. Figure 9-2 (a) and (b) are two central part palmprint sub-images extracted by this algorithm. This preprocessing technique greatly reduces the translation and rotation of palmprints captured from the same palms.

Obviously, the dimension of the OPS is $128 \times 128 = 16,384$. In biometric systems, there exist two limitations of this high dimension [83]: firstly, the recognition accuracy will decrease dramatically when the number of image classes increases. In face recognition, the typical size of training images is around 200. Second, it results in high computation complexity, especially, when the number of the classes is large. To overcome these limitations, we should reduce the resolution of the palmprint images. In face recognition, the image with 16×16 resolution is sufficient for distinguishing a human face [71, 83]. To investigate the relationship between the recognition accuracy and the resolution of palmprint images, the original images are decomposed into a Gaussian pyramid (see Figure 9-4) and the images at each level are tested. The original image is at the bottom of the pyramid (0th level) and the images at the *ith* level ($i = 1, \cdots, 5$) of the pyramid are obtained as follows: convolve the images at $(i-1)th$ level with a Gaussian kernel and then subsample the convolved images. The resolution of *ith* level image is $2^{7-i} \times 2^{7-i}$, where

$i = 0,1,\cdots,5$. At each level, six images of each palmprint class are randomly chosen as training samples to form the template and the remaining four images are used as testing samples. All of the experiments are conducted using the Microsoft Windows 2000 and Matlab 6.1 with image processing toolbox on a personal computer with an Intel Pentium III processor (900MHz).

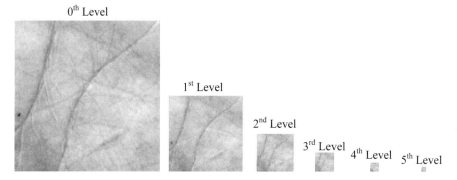

Figure 9-4. An example of Gaussian pyramid decomposition of a palmprint image.

Experimental Results

To analyze the relationship between the performance of the proposed method and image resolution, the feature vector of each testing palmprint is matched against each stored template at each level. A genuine matching is defined as the matching between the palmprints from the same palm and an imposter matching is the matching between the palmprints from different palms. A total of 360,000 (4 × 300 × 300) comparisons are performed at each level, in which 1,200 (4 × 300) comparisons are genuine matching. The genuine and impostor distributions at each pyramid level are plotted in Figure 9-5. It can be seen from this figure that there exists two peaks in the distributions at each level. One peak corresponds to genuine matching and the other corresponds to impostor matching. When the distance threshold is set as the one corresponding to the intersection of genuine and impostor distribution curves, the total error reaches the minimum, and the corresponding threshold, false acceptance rate (FAR), false rejection rate (FRR) and half total error rate (HTER, which equals (FAR + FRR)/2) [74] at each level are listed in Table 9-1. According to this table, HTERs of 0th, 1st and 2nd levels are much less than those of other levels. In other words, the palmprint image with 128 × 128, 64 × 64 and 32 × 32 resolution are more suitable for Fisherpalms based palmprint recognition than the other resolutions. Because the differences of HTERs at 0th, 1st and 2nd levels are very little (< 0.1), it is difficult to decide which level is optimal for identity recognition.

A further analysis of the images at these three levels (0th, 1st and 2nd) are made by considering their *Receiver Operating Characteristic* (ROC) curves, which plots the FAR against the FRR [74]. Figure 9-6 plots the ROC curves of 0th, 1st and 2nd levels and their corresponding equal error rate (EER, where FAR equals FRR). From this figure, the whole curve of 1st level is below that of 0th level. Hence, the palmprints at 1st level are better than those at 0th level in the proposed method. This figure also

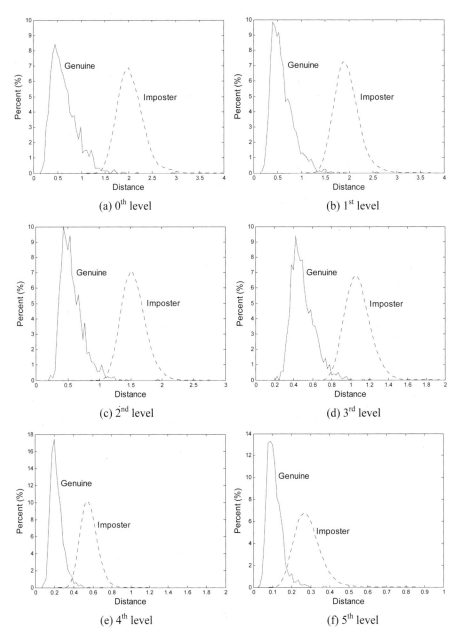

(a) 0th level

(b) 1st level

(c) 2nd level

(d) 3rd level

(e) 4th level

(f) 5th level

Figure 9-5. Genuine and imposter distributions at each pyramid level.

shows that the curve of the 2nd level is below that of the 1st level when the FAR is in the interval [0.55, 1.87] (between the dotted vertical lines in Figure 9-6, the corresponding FRR is in [0.67, 1.0]), and the curve of the 1st level is below that of the 2nd level when the FAR is smaller than 0.55 (the corresponding FRR is larger than

1.0). Therefore, images at the 2nd level (32×32 resolution) are optimal for medium security systems such as some civil systems, in which FRR should be low, while the images at the 1st level (64×64 resolution) are optimal for high security systems such as some military systems, in which FAR should be low. The EER of the 0th, 1st and 2nd levels are 1.00%, 0.95% and 0.82%, respectively (See Table 9-2).

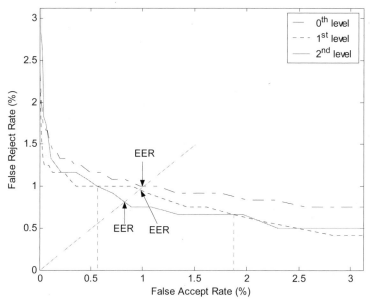

Figure 9-6. ROC curves at the 0th, 1st and 2nd levels.

Table 9-1. FAR, FRR and HTER of each pyramid level.

Pyramid level	0th	1st	2nd	3rd	4th	5th
Distance threshold	1.41	1.33	1.09	0.79	0.38	0.17
FAR (%)	0.1628	0.0814	0.1800	0.6491	1.4507	3.2302
FRR (%)	1.4167	1.2500	1.2238	2.5833	1.9167	6.3333
HTER (%)	0.7897	0.6657	0.7019	1.6162	1.6837	4.7817

Table 9-2. FAR, FRR and HTER of each pyramid level.

Pyramid level		0th	1st	2nd
Resolution		128×128	64×64	32×32
Train	Training time (seconds)	1070	47	12
One-to-one matching test	Equal error rate (%)	1.00	0.95	0.82
One-to-300 matching test	Accuracy (%)	99.20	99.25	99.75
	Testing time (seconds)	0.40	0.36	0.34

Other experiments about palmprint identification (one-to-300 matching) are also done by using the images at the 0th, 1st and 2nd levels. A nearest neighbor classifier based on Euclidean distance is employed. The identification rates of 0th, 1st and 2nd levels are 99.20%, 99.25% and 99.75%, respectively. The testing time and the identification accuracies of these levels are listed in Table 9-2. The rates and the respond time can meet the requirement of an online palmprint recognition system. The training time of these levels is also listed in Table 9-2.

Table 9-3. Comparison of different palmprint recognition methods.

Method	Duta's approach [77]	Li's algorithm [80]			Proposed method		
Database size	30 images (3 palms)	3,000 images (from 300 palms)			3,000 images (from 300 palms)		
Feature extraction	Feature points	R feature and θ feature			Fisherpalms		
Feature type	Structural feature	Statistical feature			Algebraic feature		
Image resolution	400×300	128×128	64×64	32×32	128×128	64×64	32×32
One-to-one matching accuracy rates	95.00%	96.40%	95.02%	93.24%	99.00%	99.05%	99.18%
1-to-many matching accuracy rates	Not presented	94.67%	93.00%	90.33%	99.20%	99.25%	99.75%

Performance Comparisons

Comparisons have been conducted among Duta's approach [77], Li's algorithm [80], and our proposed fisherpalms method. In Duta's approach, the lines of a palmprint were firstly extracted by directly binarizing the offline palmprint images (which were obtained by pressing an inked palm on a paper) with an interactively chosen threshold, and then some feature points and their orientation were extracted from these lines to verify the identity. 30 offline images with 400×300 resolution captured from three persons were used and 95% accuracy was obtained in the one-to-one matching test in their experiments. The feature points in this approach belonged to structural features of palmprints. It is evident that the recognition accuracies of this approach are dependent heavily on the result of the line extraction. Because of the noise, lower image resolution, and unexpected disturbance (such as the movement of hand, different lighting environment, etc) from the online palmprints which may cause lower image quality compared with those offline palmprint images. Thus it is much more difficult to extract lines from the online palmprint images. Up to now, there is no effective line extraction method for online palmprints yet. Therefore, we only use Duta's experimental results here for comparison. In Li's algorithm, the R feature and θ feature of the palmprint, which belonged to statistical features, were extracted from the frequency domain to identify different persons. R features showed the intensity of the lines of a palmprint and θ features showed the direction of these lines. However, all of these features could not reflect the spatial position of these lines since they were extracted in frequency domain. Thus their abilities to discriminate palms were not strong. Li's algorithm for 128×128, 64×64 and 32×32 resolutions has been implemented in our database. The corresponding accuracies in the one-to-one matching are 96.40%, 95.02% and 93.24%, respectively. And the accuracies in the one-to-300 matching are 94.67%, 93.00% and 90.33%, respectively. Obviously, the results of our method are much better than those of Duta's and Li's approaches because Fisherpalms is based on algebraic transforms and matrix decompositions. At the same time, the extracted algebraic features can represent the intrinsic attributions of palmprints. Table 9-3 summarizes the feature of our method compared to these two approaches with respect to database size, image resolution, feature type, feature extraction and accuracy.

9.3   Eigenpalms in Palmprint

The concept of an eigenspace has been widely used in face recognition, which shows that the extracted "eigenfaces" can effectively represent the principal components of the faces [84-85]. In this section, we will try to use similar technique to see whether it offers good characteristics for palmprint recognition. Based on the K-L (Karhunen-Loeve) transform, the original palmprint images used in training are transformed into a small set of characteristic feature images, called "eigenpalms", which are the eigenvectors of the training set. Then, feature extraction is performed by projecting a new palmprint image into the subspace spanned by the "eigenpalms".

Eigenpalms Feature Extraction
A palmprint image is described as a two-dimensional array (N by N). In the eigenspace method, this can be defined as a vector of length N^2, called a "palm vector". A central part palmprint sub-image is fixed with a resolution of 128×128, hence a vector can be obtained, which represents a single point in the 16,384-dimensional space.

Since palmprints have similar structures (usually three principal lines and wrinkles), all "palm vectors" are located in a narrow image space, thus they can be described by a relatively low dimensional space. As the most optimal orthonormal expansion for image compression, the K-L transform can represent the principal components of the distribution of the palmprints or the eigenvectors of the covariance matrix of the set of palmprint images. Those eigenvectors define the subspace of the palmprints, which are called "eigenpalms". Then, each palmprint image in the training set can be exactly represented in terms of a linear combination of the "eigenpalms".

Let the training samples of the palmprint images be $x_1, x_2, \ldots, x_M$, where M is the number of images in the training set. The average palmprint image of the training set is defined by

$$\mu = \frac{1}{M} \sum_{i=1}^{M} x_i .$$

(9-15)

The difference between each palmprint image and the average image is given by

$\varphi_i = x_i - \mu$. Then, we can obtain the covariance matrix of $\{x_i\}$ as follows:

$$C = \frac{1}{M}\sum_{i=1}^{M}(x_i - \mu)(x_i - \mu)^T ,$$

$$= \frac{1}{M}XX^T , \tag{9-16}$$

where the matrix $X = [\varphi_1\ \varphi_2\ ...\varphi_M]$. Obviously, the matrix C is of dimensions N^2 by N^2. It is evident that the eigenvectors of C can span an algebraic eigenspace and provide an optimal approximation for those training samples in terms of the mean-square error. However, determining the eigenvectors and eigenvalues of the matrix C ($C \in \mathfrak{R}^{N^2 \times N^2}$) is an intractable task for a typical image size. Therefore, we need to find an efficient method to calculate the eigenvectors and eigenvalues. It's well known that the following formula is satisfied for the matrix C,

$$Cu_k = \lambda_k u_k , \tag{9-17}$$

where u_k refers to the eigenvector of the matrix C, and λ_k is the correlative eigenvalue of matrix C.

In practice, the number of the training samples, M, is relatively small. The eigenvectors (v_k) and eigenvalues (a_k) of matrix $L = X^T X$ ($L \in \mathfrak{R}^{M \times M}$) are much easier to calculate. Therefore, we have

$$X^T X v_k = a_k v_k , \tag{9-18}$$

and we multiply each side of the Equation 9-18 by X,

$$XX^T(Xv_k) = a_k(Xv_k) . \tag{9-19}$$

Then, we can get the eigenvectors of matrix C,

$$u_k = Xv_k . \tag{9-20}$$

By using this method, the calculations are greatly reduced, where $U = \{u_k, k = 1...M\}$ denotes the basis vectors which correspond to the original palmprint images and span an algebraic subspace called unitary eigenspace of the training set. Resizing each of the eigenvectors into the image domain (N by N), we find that they are like palmprints in appearance and can represent the principal characters (especially, the principal lines) of the palmprints, which are referred as "eigenpalms". Figure 9-7 shows some of the eigenpalms derived from the samples in the training set.

Since each palmprint in the training set can be represented by an eigenvector, the number of the eigenpalms is equal to the number of the samples in the training set. However, the theory of principal component analysis states that it doesn't need to choose all of the eigenvectors as the base vectors and just those eigenvectors which correspond to the largest eigenvalues can represent the characteristic of the training set quite well. Then the M' significant eigenvectors (u'_k) with the largest associated eigenvalues are selected to be the components of the eigenpalms ($U' = \{u'_k, k = 1...M'\}$), which can span an M' dimensional subspace of all possible palmprint images. A new palmprint image is transformed into its "eigenpalms" components by the following operation,

$$f_i = U'(x_i - \mu), \qquad (i = 1, \ldots, M) \qquad (9\text{-}21)$$

where the weight of the projection f_i ($f_i \in \Re^{M' \times 1}$) refers to the standard feature vector of each person, and M' is the feature length.

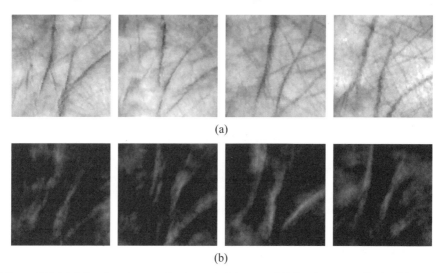

(a)

(b)

Figure 9-7. (a) Sub-palmprint samples in our training set, (b) The eigenpalms derived from the above samples.

9.4 Performance Analysis on Eigenpalms

A palmprint database of 3,056 palmprint samples is formed from the **PolyU-ONLINE-Palmprint-II**, for palmprint experiments testing. There are 382 palms from 191 persons (left and right palms). We take 8 samples from each palm to form the database. Thus, a palmprint database of 382 classes was created, which included a total of 3,056 (=191 × 2 × 8) images. The resolution of all of the original palmprint images is resized to 384 × 284 pixels, at 75 dpi, in 256 gray levels. Similar to the Fisherpalms, we use the preprocessing approach described in Chapter 6 Section 6.3 to align the palmprints.

There are four kinds of experiment schemes designed as follow: One (two, three or four) sample(s) of each person was randomly selected for training, and the other 4 samples were used for authentication, respectively. During the experiments, the features are extracted by using the proposed eigenspace method with length of 50, 100, 150 and 200. The weighted Euclidean distance is used to cluster those features [86],

$$d_k = \sum_{i=1}^{N} \frac{(f(i) - f_k(i))^2}{(\sigma_k)^2}, \qquad (9\text{-}22)$$

where f is the feature vector of the unknown palmprint, f_k and σ_k denote the kth feature vector and its standard deviation, and N is the feature length.

Based on these schemes, the matching is separately conducted and the results are reported on Table 9-4. A high recognition rate (99.149%) was achieved for the fourth scheme with feature length of 100. It can demonstrate that the feature length plays an important role on the matching process, i.e. long feature lengths lead to a high recognition rate. However, this principal only holds to a certain point as the experimental results show that the recognition rate remains unchanged, or even becomes worse, when the feature length is extended further.

Table 9-4. The testing results of the three matching schemes with different feature lengths.

Recognition rate / Training samples	Feature length			
	50	100	150	200
1	94.175%	95.550%	95.175%	93.128%
2	96.073%	97.186%	96.924%	95.942%
3	97.186%	98.429%	98.822%	97.971%
4	97.840%	99.149%	99.084%	98.691%

Analysis of Eigenpalms

A further analysis of the fourth scheme was made by calculating the standard error rates (False Acceptance Rate or FAR, and the False Rejection Rate or FRR) [18]. Obviously, for an effective method, both rates must be as low as possible, but they are actually antagonists and lowering these errors is part of an intricate balancing act. For example, if you make a system more difficult to enter for an impostor (reducing FAR), you also make the system more difficult to enter for a valid enrollee (i.e., FRR raised). This process operates in the reverse sense too. For a given system, this becomes a question of probabilities, and a company deploying such a system will generally adjust the matching threshold depending on the level of security needed. For instance, a bank needs a very secure system, so it would adjust the threshold very low to reach an FAR close to zero. However, the bank's employees will have to accept false rejections, and they may have to try several times to enter the system. The curves for the FRR and FAR of the fourth scheme are shown in Figure 9-8. When the threshold value is set to 0.71, the palmprint recognition method can achieve an ideal result with an FRR=1% and an FAR=0.03%, respectively.

Compared with the approach in [77], a set of feature points along the prominent palm lines and the associated line orientation of palmprint images were used to identify the individuals, where a matching rate about 95% was achieved. However, only 30 palmprint samples were collected from three persons for the testing. It is too small to cover the distribution of all palmprints. An average recognition rate of 91% was achieved by the technology proposed in [82], which involved a hierarchical palmprint recognition fashion. The global texture energy features were used to guide the dynamic selection for a small set of similar candidates from the database at coarse level for further processing. An interesting point based image matching was performed on the selected similar patterns at fine levels for the final confirmation.

Since multiple feature extraction methods and matching algorithms are needed, the whole process of recognition is more complex. Nevertheless, the recognition rate of eigenpalms is more efficient, as illustrated in Table 9-5.

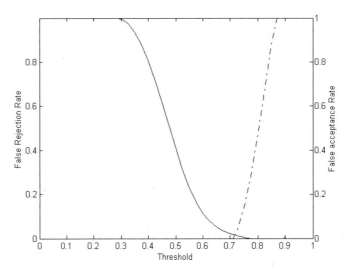

Figure 9-8. The FRR and FAR of the proposed algorithm.

Table 9-5. Comparison of different palmprint recognition methods.

Method	Duta's approach [77]	You's approach [82]	Eigenpalms
Database	30 samples	200 samples	3,056 samples
Features	Feature points	Global texture features & feature points	Eigenpalms
Recognition Rate	95%	91%	99.149%

9.5   Summary

Palmprint is an important complement of personal identification. There are many features in a palmprint, such as structural features, statistical features, and so on. In this chapter, we try to discuss another type of features, algebraic features, from palmprint. The novel palmprint recognition methods proposed in this chapter are called Fisherpalms and Eigenpalms.

Fisher's Linear Discriminant is used to project the palmprint image from the very high dimensional *original palmprint space* to the very low dimensional *Fisher palmprint space*, in which the ratio of the determinant of the between-class scatter to that of the within-class scatter is maximized. It shows that, in the Fisherpalms based palmprint recognition system, the images with resolution 32×32 are optimal for medium security biometric systems while those with resolution 64×64 are optimal for high security biometric systems. For palmprints with resolution 32×32, accuracies of 99.18% and 99.75% are obtained in one-to-one matching test and one-to-300 matching test, respectively. For palmprints with resolution 64 × 64, these accuracies are 99.05% and 99.25%. And for palmprints with resolution 128 × 128, accuracies of 99.00% and 99.20% are obtained in one-to-one matching test and one-to-300 matching test, respectively. The average testing time for the images with these resolutions in one-to-300 matching is not more than 0.4s, which is short enough for real-time palmprint recognition.

Similarly, eigenpalms method is developed by using the K-L transform algorithm. It can represent the principal components of the palmprints fairly well. The features are extracted by projecting palmprint images into an eigenpalms subspace. To assess the efficiency of our method, the weighted Euclidean distance classifier is applied. A correct recognition rate of up to 99% can be obtained using our approach.

It can be seen that both fisherpalms and eigenpalms using the algebraic features from palmprint can achieve high recognition rate.

10 TRANSFORM-BASED PALMPRINT IMAGE ANALYSIS

In this chapter, we propose a method for extracting features that converts a palmprint image from a spatial domain into a frequency domain. We first introduce a feature extraction method using a Fourier Transform. The feature extraction in the frequency domain are used as indices for the palmprint templates in the database and the searching process for the best match is leaded by a layered fashion, revealed in Section 10.1. The experimental results show that palmprint feature extraction in the frequency domain is accurate and efficient. Section 10.2 reveals another approach that uses Wavelet signatures via directional context modeling. Section 10.3 illustrates a set of statistical signatures for the characterization of a palmprint. The palmprint is transformed into the wavelet domain, the directional context of each wavelet subband is defined and computed to cluster the predominant coefficients of the principal lines and wrinkles in the associated direction. With the selected directional context values, a set of statistical signatures, which includes gravity center, density, spatial dispersity and energy, is defined to characterize the palmprint. Section 10.4 reports the identification strategies and experiments. Section 10.5 summarizes this chapter finally.

10.1 Palmprint Analysis by Fourier Transform

Fourier Transform is one of the most popular and useful transform in image processing applications [87-91]. The major applications involve image enhancement and feature extraction. Fourier Transform includes feed forward Transform and inverse Transform. The former converts an image from the spatial domain into the frequency domain and the later changes it from the frequency domain back to the spatial domain. Two-dimensional discrete Fourier Transform is defined as:

$$F(u,v) = \frac{1}{MN}\sum_{m=0}^{M-1}\sum_{n=0}^{N-1} f(m,n)\exp[-j2\pi(\frac{mu}{M}+\frac{nv}{N})], \qquad (10\text{-}1)$$

where $f(m,n)$ is an image with $M \times N$, $j = \sqrt{-1}$, $u = 0,1,...,M-1$; $v = 0,1,...,N-1$. The inverse transform is defined as:

$$f(x,y) = \sum_{m=0}^{M-1}\sum_{n=0}^{N-1} F(u,v)\exp[j2\pi(\frac{mx}{M}+\frac{ny}{N})]. \qquad (10\text{-}2)$$

In [92], the calculation on Fourier Transform is described in details. It is natural to use Fourier Transform to do the image enhancement, where a high pass filter is applied on the edge lines and a low pass filter is used to smooth the image. Figures 10-1 (c)-(e) and (f)-(h) show the results of both high pass and low pass filters to the palmprint image, respectively. It is obvious that they do not provide more values to the palmprint feature extraction except the frequency domain image as shown in Figure 10-1 (b).

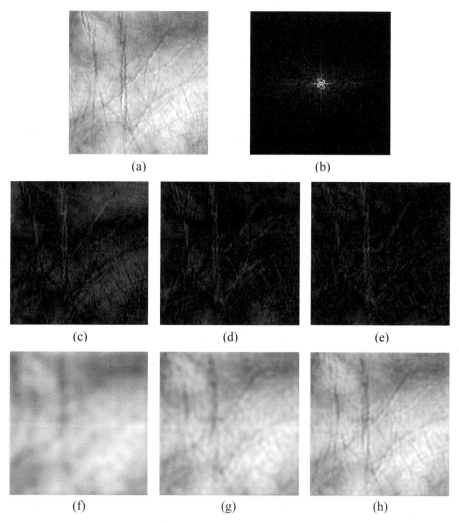

Figure 10-1. High pass and low pass filters to the palmprint images. (a) The original palmprint image, (b) The frequency domain image, (c) – (e) High pass filters (radius > 0, 1, 2), and (f) - (h) Low pass filters (radius < 10, 20, 30).

Palmprint Features Exhibited in the Frequency Domain

For palmprint identification, Fourier Transform can be used in feature extraction because there are some correspondences between palmprint features on spatial domain to its frequency domain. In general, the stronger the principal lines are on a spatial domain image, the less compact the information is on a frequency domain image. And if a palmprint image in the spatial domain has a strong line, in the frequency domain there will be more information in the line's perpendicular direction. As a result, Figure 10-2 shows three typical palmprints and their correspondence frequency domain images, where (a) is a palmprint without strong principal lines and its frequency domain image shows that the information is centralized in the center, which is the high frequency area; (b) is a palmprint with two clear and strong principal lines. Note that its frequency domain image indicates that there exists rich information on the direction perpendicular to the principal lines; and (c) is a palmprint with the full of strong wrinkles, where its frequency domain image illustrates that the information is not as centralized as that in Figure 10-2 (a).

For palmprint identification application, it is important that the similar palmprints remain resemble to each other when converted to frequency domain for the feature extraction. As an illustration, Figure 10-3 shows these three groups of palmprints and their correspondent frequency images, which are from the same palm, the similar palms and the different appearance palms.

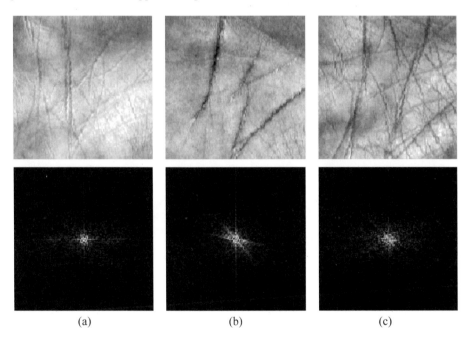

(a)   (b)   (c)

Figure 10-2. Different palmprints and their correspondence frequency domain images: (a) palmprint without strong principal lines; (b) palmprint with two clear and strong principal lines; (c) palmprint with the full of strong wrinkles.

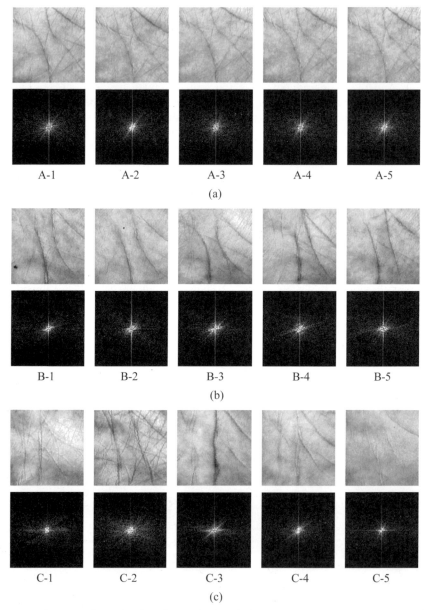

Figure 10-3. Comparison of various frequency domain images: (a) samples from the same palm, (b) samples from the similar palms, and (c) samples from the different appearance palms.

Palmprint Feature Representation

Palmprint feature representation is to describe the features in a concise and easy to compare way. If we use a polar coordination system, (r,θ), to represent the frequency domain images, the energy change tendency along r shows the wrinkles' intensity and that along θ shows the wrinkles' directions of a palmprint. Therefore, we may use a statistical method to represent palmprint's features. The image can be converted from a right angle coordination system into a polar coordination system by:

$$I'(r,\theta) = I(64 + r\cos\theta, 64 + r\sin\theta), \qquad (0 \leq r \leq 64, \quad 0 \leq \theta \leq \pi) \qquad (10\text{-}3)$$

where I is the image under right angle coordination system and I' is the image under polar coordination system.

In order to represent the intensity of principal lines, the frequency domain image is divided into some small parts by a series of circles that have the same center, as shown in Figure 10-4 (a). The energy in each ring like area is defined as:

$$R_i = \sum_{\theta=0}^{\pi} \sum_{r=8(i-1)}^{8i} I'(r,\theta), \qquad (i = 1,2,...,8) \qquad (10\text{-}4)$$

where I' is the sub image under a polar coordination system, and $R_i (i = 1,2,...,8)$ is called as R feature.

In order to represent the direction of principal lines, the frequency domain image is divided by a series of lines that go through the center of the image, as shown in Figure 10-4 (b). The energy in each fan like part is defined as:

$$\theta_i = \sum_{\theta=(i-1)}^{i} \sum_{r=0}^{64} I'(r,\theta\pi/8), \qquad (i = 1,2,...,8) \qquad (10\text{-}5)$$

where $\theta_i (i = 1,2,...,8)$ is defined as θ feature.

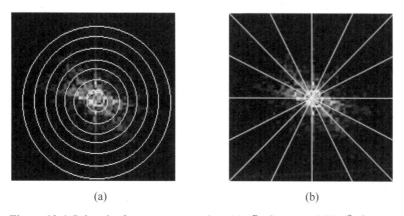

(a)   (b)

Figure 10-4. Palmprint feature representation: (a) R feature and (b) θ feature.

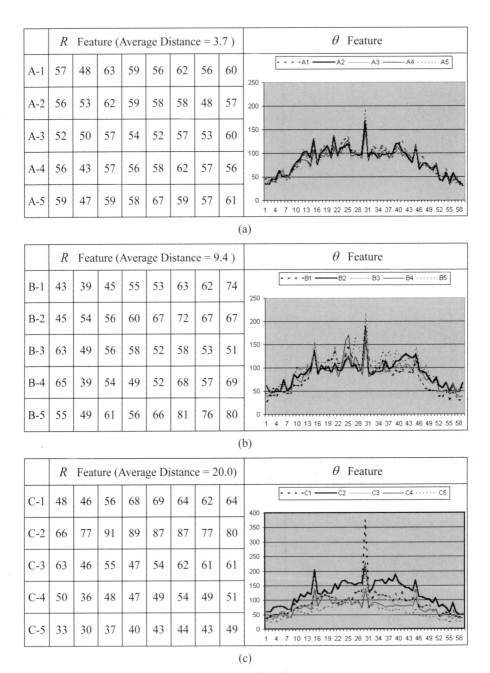

Figure 10-5. Comparison to both R and θ feature values from the samples shown in Figure 10-3, of the same palm in (a); of the similar palms in (b); and of the different appearance palms in (c).

Feature Matching
Feature matching is performed to calculate the distance between two palmprint feature sets. Since a palmprint is represented by both R features and θ features, their matching becomes to calculate the distance between R features and θ features.

Let $RX_i(i = 1,2,...,8)$ and $RY_i(i = 1,2,...,8)$ represent two R feature sets. The distance, DR_{xy}, between RX_i and RY_i is defined as:

$$DR_{xy} = \frac{1}{8}\sum_{i=1}^{8} |RX_i - RY_i|.\tag{10-6}$$

Also let $\theta X_i(i = 1,2,...,8)$ and $\theta Y_i(i = 1,2,...,8)$ represent two θ feature sets. Their distance, $D\theta_{xy}$, is defined as:

$$D\theta_{xy} = (1 - \frac{l_{xy}l_{xy}}{l_{xx}l_{yy}}) \times 100,\tag{10-7}$$

where

$$l_{xx} = \sum_{i=1}^{8}(\theta X_i - \frac{1}{8}\sum_{i=1}^{8}\theta X_i)^2,\tag{10-8}$$

$$l_{yy} = \sum_{i=1}^{8}(\theta Y_i - \frac{1}{8}\sum_{i=1}^{8}\theta Y_i)^2,\tag{10-9}$$

and

$$l_{xy} = \sum_{i=1}^{8}(\theta X_i - \frac{1}{8}\sum_{i=1}^{8}\theta X_i)(\theta Y_i - \frac{1}{8}\sum_{i=1}^{8}\theta Y_i).\tag{10-10}$$

The scope of $D\theta_{xy}$ is between 0 and 100, in which the smallest distance is 0 and the largest distance is 100. Figure 10-5 shows both values of R and θ features from the given palmprints shown in Figure 10-3. For Group A, which includes the samples from the same palm, the average distance of R features is 3.7 and the θ features are very close to each other. Group B includes the samples from the similar palms, where the average distance of R features is 9.4 and the θ features are close to each other. For Group C, which covers the samples from the palms with the different appearances, the average distance of R features is 20.0 and the θ features are far from each other. Both R feature and θ features are used in feature matching and palmprint identification.

Palmprint Identification in a Layered Fashion
Palmprint identification is to search in the database in order to find the palmprint that is from the same palm as the input one. Two key issues are involved in the searching: accuracy and efficiency. Assume that a palmprint is represented by R features and θ features, and these features are used as indices to the palmprint database. The searching is carried out in a layered fashion, where R feature is used to lead the first round of searching to obtain a candidate set, and then θ feature is applied to lead the second round of searching to produce a final output. R feature is used to

lead the first round of searching because of its fast computation on the comparisons. Figure 10-6 shows the whole identification process.

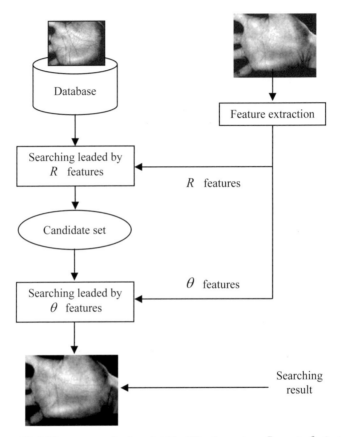

Figure 10-6. The process of palmprint identification using R and θ features.

Experimental Results and Analysis
The following experiments are achieved for testing the accuracy and efficiency of the proposed method. A database of 3,000 palmprint samples (500 persons with 6 samples per person) is formed from the **PolyU-ONLINE-Palmprint-I**, for palmprint experiments testing. These palmprints are obtained from different age group, gender, and occupations. The size of all the palmprint images is 320×240 at 75 dpi. We randomly pick up one template out of the six samples of each person to set up a 500-template database, and then use the remaining 2,500 samples as the testing set. From the algorithm, we allow the palmprint samples from the same palm to have some rotations (±30 degrees) and shifts (±50 pixels). Using the preprocessing approach described in Chapter 6 Section 6.1, the central part palmprint sub-image, 128 × 128, is obtained to represent the whole palmprint. This palmprint segmentation technique reduces the translation and rotation of the palmprints captured from the same palms.

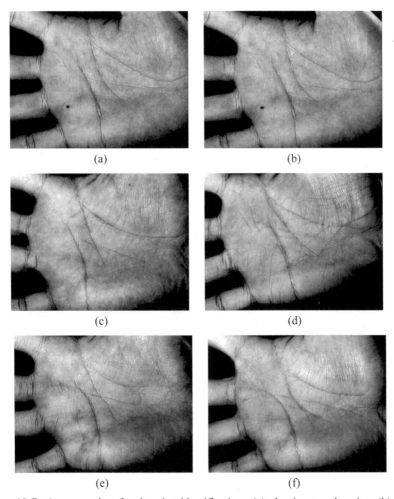

(a) (b)

(c) (d)

(e) (f)

Figure 10-7. An example of palmprint identification: (a) the input palmprint, (b) – (f) palmprints of the shortest distances with (a) in the ascending order.

The testing process is that for each sample in the testing set, find out the template in the 500-template database, which is come from the same palm. The output for a query may be correct or not correct. We count the number of correct answers and record the response time to evaluate the accuracy rate and efficiency of the proposed method. Table 10-1 shows the testing results on our database with 500 templates, where the identification accuracy is 95.48%, the shortest response time is about 1.2 seconds, the longest time is about 3.7 seconds and the average time is about 2 seconds in the layered fashion searching. Figures 10-7 illustrated the identification process using this method, where (a) is an input sample, and (b)-(f) are the templates selected by the first round searching ordered by the shortest distance. Figure 10-7 (b) is the final output decided by the second round searching.

Table 10-1. The results of the palmprint performance testing.

Templates in the database	500
Attempts in testing	2500
Correct answers	2387
Identification rate	95.48%
Average response time (s)	2

10.2   Wavelet Signatures via Directional Context Modeling

Statistical approaches are the most intensively studied and used frameworks in the fields of pattern recognition and feature extraction [104]. In these approaches, the pattern is represented by a feature vector of *n*-dimensions. The effectiveness of a method is determined by how well the different patterns can be separated by the representation vector. In the last decades, wavelet-based statistical feature extraction approaches [105-109] have been reported with good results. Wavelet transform [94-97] is a time-frequency analysis and it endows the traditional feature extraction methods with a multiresolution and multiscale framework. As the orthogonal wavelet transform (OWT) is translation variant, which is a limitation in the texture analysis, we employ the overcomplete wavelet expansion (OWE), which has the property of translation invariant, as shown in our proposed method. No subsampling occurs in the decomposition of OWE and at each scale the wavelet coefficients are of the same size.

A palmprint can be considered as an ordinary grayscale image, to which the traditional statistical feature analysis methods can be applied. However, a palmprint has many particular characteristics other than an ordinary image. The locations, shapes and distributions of the principal lines and wrinkles in a palmprint have conveyed redundant information to identify a person uniquely. From these features, a set of statistical signatures will be derived to characterize a palmprint. We decompose the palmprint into several scales by OWE. At each scale and each subband, the predominant coefficients of the principal lines and wrinkles are clustered. Some statistical signatures include gravity center, density, spatial dispersity, and energy, can be defined to measure the characteristics of the input palmprint. A hierarchical classification and identification scheme is sequentially yielded based on these signatures. The presented scheme exploits the features of principal lines and wrinkles sufficiently and achieves satisfactory results. Compared with the line segments based or interesting points based matching schemes, the proposed scheme uses a smaller amount of data signatures, and yields an efficient classification strategy with higher accuracy rate.

Structure of the Model

Wavelet transform (WT) represents a function f as a linear combination of elementary atoms or building blocks. A detailed description of the theory of wavelet and its relationship with signal processing can be found in Daubechies [94], Mallat *et al.* [95-96] and Vetterli *et al.* [97]. Denote by $\psi_{m,n}$ the dyadic dilation and translation of a mother wavelet ψ with $m, n \in Z$

$$\psi_{m,n}(t) = 2^{-m/2}\psi(2^{-m}t - n). \tag{10-11}$$

Then f can be written as:

$$f = \sum c_{m,n}(f)\psi_{m,n}. \tag{10-12}$$

For orthonormal wavelet bases, there is:

$$c_{m,n}(f) = \langle f, \psi_{m,n} \rangle = \int f(t)\psi_{m,n}(t)dt, \tag{10-13}$$

Where $\langle \cdot, \cdot \rangle$ is the inner product in $L_2(\Re)$. For bi-orthogonal wavelet bases, we have:

$$c_{m,n}(f) = \langle f, \tilde{\psi}_{m,n} \rangle, \tag{10-14}$$

where $\tilde{\psi}_{m,n}$ is the dual wavelet of $\psi_{m,n}$. Except for the Haar wavelet, all the compactly supported orthogonal wavelets are not (anti-)symmetrical [94], which is a very important property in signal processing. Compactly supported bi-orthogonal wavelet trades the orthogonality for the symmetric property.

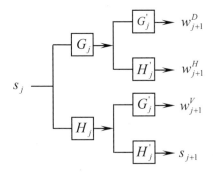

Figure 10-8. One stage decomposition of the 2-D overcomplete wavelet expansion (OWE).

Orthogonal wavelet transform (OWT) is variant with the translation of the input signal due to the subsampling in the decomposition process. This limits its efficiency in some signal processing applications, such as denoising [102] and texture analysis [108-109]. The feature extraction scheme presented in this chapter is implemented with the overcomplete wavelet expansion (OWE), whose one stage decomposition structure is shown in Figure 10-8. The right arrows indicate subsampling. H_j and G_j are the lowpass and highpass analytic filters. H'_j and G'_j are the transpose filters of them. H_j is interpolated by putting $(2^{j-1} - 1)$ zeros between each of the coefficients of H_0, so does for G_j. The output of the OWE is invariant with the translation of the input signal. No subsampling occurs in the decomposition but the analytic and synthetic filters vary in each stage. With OWE, the bandwidth decrease is accomplished by zeros padding of filters instead of subsampling of wavelet coefficients. The highpass wavelet coefficients of OWE are in three directions:

horizontal, vertical and diagonal at each scale, and they are denoted by w_j^H, w_j^V and w_j^D. The size of them is the same as the input image.

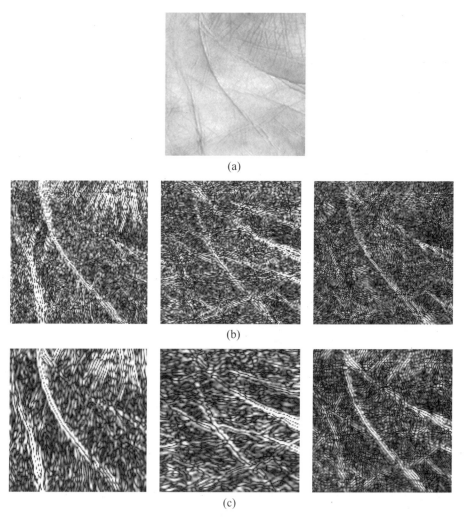

(a)

(b)

(c)

Figure 10-9. Original palmprint image in (a). The wavelet coefficients in the three directions from left to right (Horizontal, Vertical, Diagonal) at the first scale in (b), while the second scale in (c).

Figure 10-9 (a) shows a preprocessed palmprint image by the approach given in Chapter 6 Section 6.1, and (b)-(c) show the two levels OWE of it. (For visual convenience, the pixels' magnitudes are rescaled.) Figure 10-9 (b) shows the wavelet coefficients in the three directions from left to right (Horizontal, Vertical, Diagonal) at the first scale, and (c) shows them at the second scale.

10.3   Characterization of Palmprint by Wavelet Signatures

Directional Context Modeling of the Wavelet Coefficients
From Figure 10-9 we can see that the principal lines and the thicker wrinkles are enhanced in the wavelet coefficients. Generally speaking, the edges in horizontal direction are detected in w_j^H, and the edges in vertical and diagonal directions are detected in w_j^V and w_j^D. However, these edges are distributed rather dispersedly. In each subband, it is expected that we could cluster the interested edges with similar structures together, particularly, grouping the most important horizontally distributed edges in w_j^H, and the vertically and diagonally distributed edges in w_j^V and w_j^D.

The context modeling technique, which was widely used in coding [99-100] and denoising [98] to differentiate and to gather pixels with some similarities (but not necessarily spatially adjacent), is a good choice for classification. The context value of a given coefficient is defined as a function of its neighbors. The weighted average of its adjacent pixels is often employed. By computing the context of each wavelet coefficient, the pixels with similar characteristics, whose context values fall into a specified field, can be collected together.

The widths of the principal lines and some heavy wrinkles are much greater than those of other features in a palmprint. In Figure 10-10 (a)-(d), zoom-in images of the typical edge structures of the principal lines in horizontal, vertical and diagonal directions are illustrated. It shows that the edges in Figure 10-10 (a) are of considerable width across the horizontal pixels, while the edges in (b) possess a sizable width across vertical pixels. The diagonal edges in Figure 10-10 (c) and (d) have shorter widths compared with those in (a) and (b), while they propagate consecutively in the diagonal direction. These properties can be exploited to determine the interested feature structures in the associated subband.

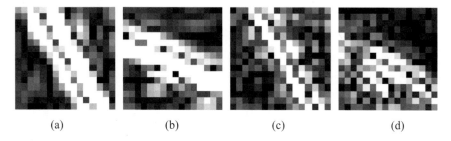

(a)   (b)   (c)   (d)

Figure 10-10. Zoom-in images of the edge structures in the three directions of the wavelet coefficients. (a) Horizontal edge structure; (b) vertical edge structure; diagonal edge structure in the same position of that in (a) and (b) are shown in (c) and (d), respectively.

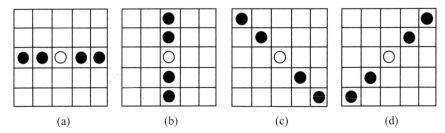

Figure 10-11. Four neighborhoods points (black points) are employed in the calculation of the directional context values: (a) horizontal direction; (b) vertical direction; (c) diagonal direction for right hand palmprint; (d) diagonal direction for left hand palmprint.

It is observed that if a horizontal edge point, which is produced by principal lines or heavy wrinkles, occurs at $w_j^H(n,m)$, its left and right neighborhoods are also likely to be horizontal edge points. Similarly, if a vertical edge point produced by principal lines or heavy wrinkles occurs at $w_j^V(n,m)$, its upside and downside neighborhoods are also likely to be vertical edge points. As for the diagonal edges, if the palmprint image is captured with the right hand, the diagonal edge directions of the principal lines are basically along $-\pi/4$; if the palmprint is of the left hand, the diagonal edges produced by the principal lines are basically along $\pi/4$. With these observations, we then defined the directional context of w_j^*, $* = H,V,D$, and cluster the interested edges by their context values.

Referring to Figure 10-11 (a)-(d), for each wavelet coefficient $w_j^*(n,m)$ in a subband, denote by $w_{j,(n,m)}^*(1)$ - $w_{j,(n,m)}^*(4)$, the absolute values of its four neighborhood elements. We define:

$$\bar{u}_{j,(n,m)}^* = \left| w_{j,(n,m)}^*(1) \quad \cdots \quad w_{j,(n,m)}^*(4) \quad \left| w_{j+1}^*(n,m) \right| \right| \tag{10-15}$$

as the associated vector of $w_j^*(n,m)$. The absolute values rather than the original values are used in the context calculation because the orthogonal wavelet coefficients are weakly correlated, but the absolute values of neighboring coefficients are correlated [103]. Thus the average of absolute values would yield more meaningful information than the original values. The parent coefficient $w_{j+1}^*(n,m)$ at the adjacent coarse scale is also introduced in the vector $\bar{u}_{j,(n,m)}^*$ since the wavelet coefficient dependencies do not only exist within each scale but also between scales [101]. If $w_j^*(n,m)$ is a significant coefficient at a scale, its parent coefficients at the coarser scales are much likely to be significant too. Define:

$$\bar{U}_j^* = \begin{bmatrix} \bar{u}_{j,(1,1)}^* \\ \bar{u}_{j,(1,2)}^* \\ \vdots \\ \bar{u}_{j,(N,M)}^* \end{bmatrix} \quad \text{and} \quad \bar{Y}_j^* = \begin{bmatrix} w_j^*(1,1) \\ w_j^*(1,2) \\ \vdots \\ w_j^*(N,M) \end{bmatrix}, \tag{10-16}$$

where N and M are the numbers of total rows and columns of subband w_j^*. $\bar{U}_j^*$ is a $N \cdot M \times 5$ matrix and $\bar{Y}_j^*$ is a $N \cdot M$ length column vector. The directional context value of $w_j^*(n,m)$ is defined as:

$$c_j^*(m,n) = \bar{u}_{j,(n,m)}^* \bar{h}_j^*, \tag{10-17}$$

where $\bar{h}_j^*$ is a 5×1 weighted vector and it can be calculated by the least square estimate:

$$\bar{h}_j^* = ((\bar{U}_j^*)^T \bar{U}_j^*)^{-1} (\bar{U}_j^*)^T |\bar{Y}_j^*|. \tag{10-18}$$

It should be noted that because the absolute values of the wavelet coefficients are used in the calculation, the context values are mostly positive.

With context modeling, the coefficients of similar natures can be well clustered. Similar structures have approximate context values, and by sorting the context values c_j^* in ascending order, the wavelet coefficients $w_j^*(n,m)$ could be classified into several groups. Since the interested horizontal edge structures produced by the principal lines or heavy wrinkles have high magnitudes, consequently their directional context values are predominant in c_j^*. We collect the most significant L_j coefficients in c_j^* as the edge points in the associated direction. Considering that the edge structures will be more and more enhanced with the increase of scale number, the number L_j should be increased accordingly, and we set it to be proportional to the scale parameter j:

$$L_j = L * j, \tag{10-19}$$

where L is a constant number preset.

By the above directional context modeling, the horizontal, vertical and diagonal edge structures can be more accurately determined than directly thresholding the wavelet coefficients $w_j^*(n,m)$. In Figures 10-12 (a)-(c), the most significant 200 points of the wavelet subband w_1^H, w_1^V and w_1^D obtained from Figure 10-9 (b) are shown respectively. The greatest 200 coefficients in the corresponding context matrix c_1^H, c_1^V and c_1^D are illustrated in Figures 10-12 (d)-(f), respectively. Obviously, the edges in Figures 10-12 (d)-(f) are more concentrative around the principal lines, while the edges in (a)-(c) are a little more dispersive. Too much information on minor wrinkle features are preserved in Figures 10-12 (a)-(c) while being eliminated in (d)-(f).

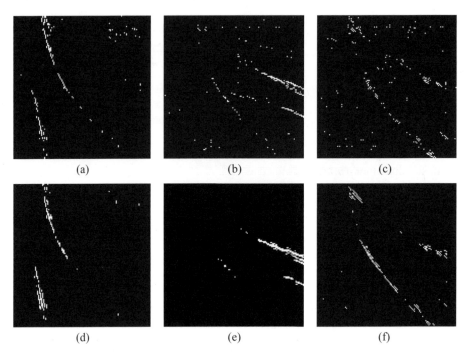

Figure 10-12. (a)-(c) show the most 200 significant wavelet coefficients in w_1^H, w_1^V and w_1^D respectively; (d)-(f) show the most 200 significant context coefficients in c_1^H, c_1^V and c_1^D respectively.

Characterizing the Palmprint by Context-Based Wavelet Signatures

Various wavelet signatures have been proposed in [105-109], and the popularly used ones are energy, histogram, and co-occurrence signatures. As for palmprint images, these signatures neither completely exploit their features nor sufficiently characterize the palmprints. Some signatures may even be inappropriate for palmprints classification. For example, the wavelet histogram signature based on detailed wavelet coefficients can be modeled by generalized Gaussian distributions (GGD) [95, 98], which has three parameters (i.e. the signatures). But we found that unlike other natural images, the histogram of the detail wavelet coefficients for a palmprint can be well approximated by Gaussian distribution, which is the degenerateness of GGD. Thus the histogram signatures are of very limited efficiency and should not be employed in the classification and identification of palmprints.

For the context value matrix c_j^* of each subband w_j^*, $* = H, V, D$, denote by t_j^* the L_jth great value in it. We can define:

$$\bar{c}_j^*(n,m) = \begin{cases} c_j^*(n,m) & \text{if } c_j^*(n,m) \geq t_j^* \\ 0 & \text{if } c_j^*(n,m) < t_j^* \end{cases}, \qquad (10\text{-}20)$$

where $\bar{c}_j^*$ is the interested directional context matrix (IDCM) of w_j^*. In Figure 10-13, some typical palmprints and their IDCMs at the second sale, $\bar{c}_2^*$, $* = H, V, D$, are shown.

It is important for us to define a set of context-based statistical signatures to characterize a palmprint completely, i.e.,

1) The signature of average gravity center;
2) The signature of density;
3) The signature of spatial dispersity, and
4) The signature of energy.

1) The signature of average gravity center

If we view $\bar{c}_j^*$ as an object and the magnitude of each point $\bar{c}_j^*(n,m)$ as the local mass in the position, then there exists a gravity center of $\bar{c}_j^*$, which could measure globally the distribution of its mass. Denote $\left(x_j^*, y_j^*\right)$ as the gravity center of $\bar{c}_j^*$, the components of the pair is defined as:

$$
\begin{cases}
x_j^* = \dfrac{1}{G_j^*} \sum_{n=1}^{N} \sum_{m=1}^{M} m \cdot \bar{c}_j^*(n,m) \\
y_j^* = \dfrac{1}{G_j^*} \sum_{n=1}^{N} \sum_{m=1}^{M} n \cdot \bar{c}_j^*(n,m),
\end{cases}
\tag{10-21}
$$

where G_j^* is the mass of $\bar{c}_j^*$, i.e.,

$$
G_j^* = \sum_{n=1}^{N} \sum_{m=1}^{M} \bar{c}_j^*(n,m).
\tag{10-22}
$$

For a fixed direction (horizontal, vertical or diagonal), the shapes of the IDCMs are similar to each other across scales, and the gravity centers of $\bar{c}_j^*$ are of close values along different scales with some variations. The signature of the average gravity center (SAGV) of a IDCM $\bar{c}_j^*$, denoted by $\left(\bar{x}^*, \bar{y}^*\right)$, is defined as the mean of gravity center sequences $\left(x_1^* \quad \cdots \quad x_J^*\right)$ and $\left(y_1^* \quad \cdots \quad y_J^*\right)$:

$$
\begin{cases}
\bar{x}^* = \dfrac{1}{J} \sum_{j=1}^{J} x_j^* \\
\bar{y}^* = \dfrac{1}{J} \sum_{j=1}^{J} y_j^*,
\end{cases}
\tag{10-23}
$$

where J is the total number of wavelet decomposition level.

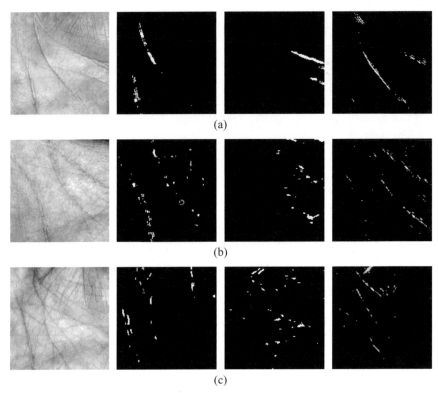

Figure 10-13. Some typical palmprint images and their interested directional context matrices (IDCM) at the second wavelet scale in the horizontal, vertical and diagonal directions. From left to right are the original image, horizontal IDCM, vertical IDCM, and diagonal IDCM, where: (a) palmprint 1; (b) palmprint 2; (c) palmprint 3.

2) The signature of density

For palmprints with deep, concentrative principal lines and weak wrinkles (for example, the first palmprint in Figure 10-13), the points in the associated IDCMs are mostly concentrated around the principal lines. For palmprints with shallow, sparse principal lines and relatively strong wrinkles, the points in the associated IDCMs are more sparsely distributed. Referring to Figure 10-13, intuitively the IDCM of the first palmprint is more compact, and the IDCMs of the other palmprints are looser.

The density of an IDCM could be defined to characterize the above-mentioned feature of a palmprint. For the ith non-zero coefficient in $\overline{c}_j^*$, denote it as $\overline{c}_j^*(i)$, and let $\Lambda_{j,(i)}^*$ be a square window centered at $\overline{c}_j^*(i)$ and with a proper size l. Denote by $K_{j,(i)}^*$ the number of non-zero points in $\Lambda_{j,(i)}^*$, and then the signature of density (SD) of IDCM $\overline{c}_j^*$ is defined as:

$$D_j^* = \frac{1}{L_j} \sum_{i=1}^{L_j} K_{j,(i)}^* . \tag{10-24}$$

The density signature is proportional to the compactness of points in the IDCM. For example, set l to be 5, the D_1^*'s of the first palmprint in Figure 10-13 is calculated to be 8, 10 and 6 respectively in the horizontal, vertical and diagonal directions, while those of the second palmprint are 5.7, 6 and 3, respectively.

3) The signature of spatial dispersity

The SD measures the compactness of the points in an IDCM, but it does not exploit the spatial distribution of the points. From Figure 10-13, we can see that although the last two IDCMs have relatively close compactnesses, they have much difference in the spatial distribution of the points.

For a typical palmprint, approximately the heart line is vertical, the life line is horizontal, and the head line is diagonal. This is also well reflected in the IDCMs. Referring to Figure 10-13, suppose we project the vertical IDCM of the first palmprint into y-coordinate, it can be imagined that the resulted projection will concentrate mainly in two local areas. But if we project the vertical IDCM of the last palmprint into y-coordinate, it can be imagined that the obtained projection will be more evenly distributed.

We first assign an associated projection vector (APV) to each IDCM. For the horizontal IDCM $\bar{c}_j^H$, which is projected into the x-coordinate, the APV is defined as:

$$\bar{p}_j^H(m) = \sum_{n=1}^{N} \bar{c}_j^H(n,m), \quad m = 1,2,\ldots,M. \tag{10-25}$$

The vertical IDCM $\bar{c}_j^V$ is projected into the y-coordinate, and its APV is defined as:

$$\bar{p}_j^V(n) = \sum_{m=1}^{M} \bar{c}_j^V(n,m), \quad n = 1,2,\ldots,N. \tag{10-26}$$

As for the diagonal IDCM $\bar{c}_j^D$, it should be projected along the direction of $-\pi/4$ for the right hand, and along the direction of $\pi/4$ for the left hand. Here we consider the right hand and the result of the left hand can be derived similarly. The APV of $\bar{c}_j^D$, denoted by $\bar{p}_j^D(k)$, will have K elements, where $K = N + M - 1$. Suppose $M \geq N$, $\bar{p}_j^D(k)$ is defined as:

$$
\bar{p}_j^D(k) = \begin{cases} \sum\limits_{m=1}^{k} \bar{c}_j^D(N-k+m,m) & k = 1,2,\ldots,N \\ \sum\limits_{m=k-N+1}^{k} \bar{c}_j^D(N-k+m,m) & k = N+1,\ldots,M \\ \sum\limits_{m=k-N+1}^{M} \bar{c}_j^D(N-k+m,m) & k = M+1,\ldots,K. \end{cases} \tag{10-27}
$$

The result when $M < N$ can be derived similarly. The APV is then normalized as:

$$
\bar{\bar{p}}_j^* = \frac{\bar{p}_j^*}{\sum\limits_i \bar{p}_j^*(i)}. \tag{10-28}
$$

We call $\bar{\bar{p}}_j^*$ the normalized associated projection vector (NAPV) of $\bar{c}_j^*$. The signature of spatial disperse (SSD) is defined as the reciprocal of the standard deviation of $\bar{\bar{p}}_j^*$:

$$
s_j^* = \frac{1}{\sqrt{\dfrac{1}{I}\sum\limits_{i=1}^{I}\left(\bar{\bar{p}}_j^*(i) - \overline{\bar{\bar{p}}_j^*}\right)^2}}, \tag{10-29}
$$

where

$$
\overline{\bar{\bar{p}}_j^*} = \frac{1}{I}\sum_{i=1}^{I}\bar{\bar{p}}_j^*(i), \tag{10-30}
$$

is the mean of the NAPV $\bar{\bar{p}}_j^*$.

The NAPVs $\bar{\bar{p}}_j^*$ of the IDCMs $\bar{c}_j^*$ in Figure 10-13 are plotted in Figure 10-14. If an NAPV of a palmprint is locally concentrated, such as the NAPVs in Figure 10-14 (a), its standard deviation will be high, and then the SSD will be of small value, which implies that the IDCM is narrowly distributed in the corresponding direction. Contrarily, the more evenly an NAPV spread, the greater its SSD will be, which implies that the IDCM is widely distributed in the direction. The SSDs of the three horizontal NAPVs in Figures 10-14 are 71.7, 109.3 and 75.5, respectively. The SSDs in the vertical direction are 82.8, 86.0 and 102.8, respectively, and the SSDs in the diagonal direction are 100, 110.4 and 131.5, respectively.

4) The signature of energy

The energy signature is one of the most popularly used wavelet signatures. In many publications [105-107, 109], the energy signature has been successfully employed in texture classification. Chang et al. [105] characterized a class of middle frequency dominated textures only by their energy signatures with a tree structured wavelet packet transform. And in [109], it is shown that better performance can be achieved by combing the wavelet energy signature with the wavelet histogram and the co-occurrence signatures.

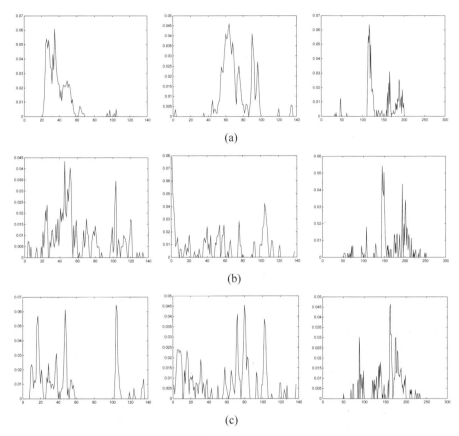

(a)

(b)

(c)

Figure 10-14. The normalized associated projection vectors (NAPV) of the ICDMs in Figure 10-13. From left to right: horizontal, vertical and diagonal NAPVs for palmprint 1 in (a), for palmprint 2 in (b) and for palmprint 3 in (c).

Traditionally, the energy signature is defined as the global power of each wavelet subband w_j^*,

$$e_j^* = \frac{1}{N \cdot M} \sum_{n=1}^{N} \sum_{m=1}^{M} \left(w_j^*(n,m)\right)^2 . \qquad (10\text{-}31)$$

Since the wavelet coefficients within each subband are well modeled by the GGD, which is zero mean, the energy signature e_j^* is truly the variance of the wavelet coefficients, and the sequence $\left\{e_j^*\right\}_{j=1,2,\dots,J}$ reflects the distribution of energy along the frequency axis over scales for a fixed direction. However, the energy signature defined above represents the variation of the global palmprint image, but not the structures we interested. We can define

$$\overline{w}_j^*(n,m) = \begin{cases} w_j^*(n,m) & \text{if} \quad c_j^*(n,m) \geq t_j^* \\ 0 & \text{if} \quad c_j^*(n,m) < t_j^* \end{cases}, \quad * = H,V,D \qquad (10\text{-}32)$$

then $\overline{w}_j^*$ is called the associated wavelet coefficients matrix (AWCM) of IDCM $\overline{c}_j^*$.

Instead of computing the total power of w_j^*, we define the signature of energy (SE) as:

$$\overline{e}_j^* = \frac{1}{L_j} \sum_{n=1}^{N} \sum_{m=1}^{M} \left(\overline{w}_j^*(n,m) \right)^2 . \qquad (3\text{-}33)$$

Similar to w_j^*, $\overline{w}_j^*$ is also nearly zero-mean, so SE $\overline{e}_j^*$ is approximately the variance of the nonzero elements in $\overline{w}_j^*$. Compared with e_j^* defined in Equation 10-31, $\overline{e}_j^*$ reflects more accurately the variation of the interested structures.

10.4   Identification Strategies and Experiments

The Classification and Identification Strategies

With the context-based statistical signatures defined and tailored for palmprint images in the last section, a hierarchical classification scheme of palmprint is directly yielded. These signatures are summarized in Table 10-2. Suppose the palmprint images are decomposed into J wavelet scales, thus for each palmprint image, it has 3 pairs of SAGC signatures, $3 \cdot J$ SD signatures, $3 \cdot J$ SSD signatures and $3 \cdot J$ SE signatures. Generally speaking, these statistical signatures can describe a palmprint accurately and then can be used to identify a person uniquely. Compared with the line segments or interesting points based palmprint verification schemes, the data storage requirement of the proposed approach is particularly lower.

The hierarchical classification and identification scheme is illustrated in Figure 10-15. The SAGC signatures are used for the first level classification. While the SD and SE signatures are used for the second level classification. If an input palmprint belongs to a subclass, then it is compared with all the palmprints in that subclass for identification.

Since different signatures have different metrical units, their compared errors could not be added up directly, so the relative error is computed in the identification process. Denote by $E_{r,(i)}^k$, $i = 1,2,\ldots,4$, the total relative error (TRE) over scales and directions between the input palmprint and the kth palmprint in the subclass, where subscript " i " represents signature SAGC, SD, SSD and SE respectively. The weighted distance (WD) between the input palmprint and the kth palmprint in the subclass is calculated by:

$$D_w^k = \sum_{i=1}^{4} \lambda_i E_{r,(i)}^k , \qquad (10\text{-}34)$$

where $\{\lambda_i\}_{i=1,2,\ldots,4}$ is a preset weighted sequence, which is determined by the reliability and stability of the signatures. Denote:

$$D_w = \min\left\{D_w^k\big|_{k=1,2,\ldots}\right\}. \tag{10-35}$$

If $D_w \leq T$, where T is the preset threshold, then the input palmprint is verified as the one whose WD from the input palmprint is equal to D_w. Otherwise, the input palmprint is judged out of the database.

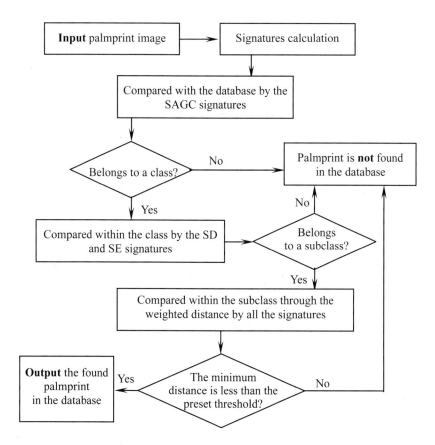

Figure 10-15. The hierarchical classification and identification strategy of palmprint.

Table 10-2. Summary of the wavelet signatures defined and tailored for palmprint images, where J is the total decomposition scale of wavelet transform.

Signature	Total number	Description
Signature of average gravity center (SAGC)	3 pairs	The average of the gravity centers of the interested directional context matrices (IDCM) over all scales. They measure the global position of the interested structures of principal lines and heavy wrinkles in a palmprint.
Signature of density (SD)	$3J$	The average number of the nonzero elements that exist in the square window centered at a nonzero element in an IDCM. They measure the compactness of the points in an IDCM.
Signature of spatial disperse (SSD)	$3J$	The reciprocal of the standard deviation of the normalized associated projection vector (NAPV) of an IDCM. They measure the spatial distribution of an IDCM in the corresponding direction.
Signature of energy (SE)	$3J$	The mean energy of the nonzero elements in the associated wavelet coefficients matrix (AWCM) of an IDCM. An SE measures the variance of the interested structures in an AWCM, and a sequence of SEs reflects the distribution of energy along the frequency axis over wavelet scales.

Experiments

A palmprint database of 200 palmprint samples (50 persons with 1 sample from 1st time captured while 3 samples from 2nd time captured) is formed from the **PolyU-ONLINE-Palmprint-II**, for palmprint experiments testing. The resolution of all of the original palmprint images is resized to 384×284 pixels, at 75 dpi. By using the preprocessing approach described in Chapter 6 Section 6.3 to align the palmprints. In this technique, the tangent of two finger holes are computed and used to align the palmprint. The central part palmprint sub-image, 128×128, is obtained to represent the whole palmprint. Such preprocessing greatly reduces the translation and rotation of the palmprints captured from the same palms. Among the 200 palmprints, 50 samples were used to calculate the signatures and the other 150 samples were used for identification. Since we are only using a small database for the experiment testing, we only performed the first level classification by the SAGC signatures and then identified the input palmprint by the D_w.

The wavelet employed in the experiments is bi-orthogonal wavelet $CDF(1,3)$ constructed in [93]. Each of the palmprint images (whose sizes is 150×150) was decomposed into three scales, i.e., $J = 3$. The distribution of the SAGCs in the Euclidean space is shown in Figure 10-16. For each direction, the palmprints were divided into two categories according to the position of their gravity centers (the two categories should have some overlap near the boundary, which is illustrated in Figure 10-16), so the 50 palmprints were classified into $2^3 = 8$ categories, and each of the categories has similar number of palmprints. In general, the SAGCs are the most reliable signatures. In the identification step, the weighted sequence was set as $\lambda_i = \{1.2, 1, 0.8, 1\}$. The identification results are summarized in Table 10-3. The corrected identification rate (CIR) is 98%, and the false identification rate (FIR, or FAR) is 0.67%, while the false rejection rate (FRR) is 1.33%. Although the SSD signature has the smallest weight, it is helpful in improving the CIR. If we set its weight $\lambda_3 = 0$, then the CIR decreases to 96.67%, and the FAR and FRR increase to 1.33% and 2%, respectively.

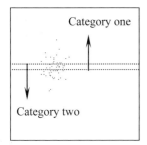 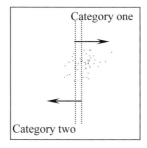 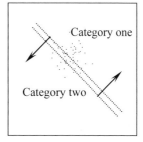

Figure 10-16. The spatial distribution of the signatures of average gravity center (SAGC) of the 50 palmprints in our database: (a) Horizontal direction, (b) vertical direction, and (c) diagonal direction.

Table 10-3. The palmprints identification results.

	Number	Percentage
Total palprints	150	100%
Correctly recognized	147	98%
Falsely recognized	1	0.67%
Falsely rejected	2	1.33%

10.5 Summary

In this chapter, a novel feature extraction method by converting a palmprint image from a spatial domain to a frequency domain is proposed. The first part of this chapter introduced the feature extraction method using Fourier Transform. It can be seen that the similar palmprints remain resemble to each other when converted to frequency domain while different palmprints are separated far from each other. Experimental results show that the identification accuracy is 95.48%.

Then, another approach using Wavelet signatures is reported. It is developed by characterizing a palmprint with a set of statistical signatures. The palmprint is transformed into the wavelet domain, and then cluster the predominant structures by context modeling according to the appearances of the principal lines in each subband. With the interested context image, we characterize an input palmprint by a set of tailor-made statistical signatures. Some of the signatures are used to classify the palmprints hierarchically, and all the signatures are used to calculate the weighted distances between the palmprints and the database. The proposed scheme gives a sufficient statistical description of the principal lines and heavy wrinkles, which convey considerable information for individual identification. Two hundreds palmprint images from fifty persons were used to perform the experiments. The fifty individuals were classified into eight categories and the corrected recognition rate is 98%. This means that the result is encouraging.

11 ONLINE PALMPRINT CLASSIFICATION

In recent years, palmprints have been investigated extensively in automated personal authentication. Duta [77] extracted some palm lines feature points from offline palmprint images for verification. Zhang [65] used 2-D Gabor filters to extract texture features from low resolution palmprint images and employed these features to implement a highly accurate online palmprint recognition system. Han [66] used Sobel and morphological operations to extract line-like features from palmprints. Kumar [67] integrated line-like features and hand geometric features for personal verification. All of these methods require that the input palmprint should be matched against a large number of palmprints in a database, which is very time consuming. To reduce the search time and computational complexity, it is desirable to classify palmprints into several categories so that the input palmprint need match only the palmprints in its corresponding category, a subset of palmprints in the database.

In Chapter 4 Section 4.3, we used the orientation property of the ridges on palms to classify offline palmprints into six categories. Obviously, this classification method is unsuitable for low resolution online palmprints because it is impossible to obtain the orientation of the ridges from these low resolution images. As our first attempt at low resolution online palmprints classification, we classify palmprints by taking into account their most visible and stable features, the principal lines. In this chapter, we first describe the palmprint preprocessing scheme. In Section 11.2 we disclose the principal lines extraction methods. Section 11.3 reports the palmprint classification strategy. Section 11.4 discusses the experimental results of the proposed method. Finally, Section 11.5 summarizes this chapter.

11.1 Online Palmprint Preprocessing

Notations and Definitions
Because there are many lines in a palmprint, it is very difficult – without explicit definitions – to distinguish principal lines from mere wrinkles. By the way, this definition is established which is not the same as the definitions on the offline palmprint images. When people discriminate between principal lines and wrinkles, the position and thickness of the lines play a key role. Likewise, we define the principal lines according to their positions and thickness.

Most palmprints show three principal lines: heart line, head line and life line, as shown in Figure 11-1 (a). To determine the positions of principal lines, we first define some points and straight lines in a palmprint. Figure 11-1 (b) illustrates these points and straight lines. Points A, B, C, F and G are the root points of the thumb, index finger and little finger. Points D and E are the midpoints of the root line of the middle finger and the ring finger. Line *BG* is the line passing through Points *B* and *G*. Line *AH* is the straight line parallel with Line *BG*, intersects with the palm boundary at Point *H*. Line *AB* is the line passing through Points *A* and *B*. Line *CL* is the straight line parallel with Line *AB* and intersecting with Lines *BG* and *AH* at Points *M* and *L*, respectively. Line *GH* passes through Points *G* and *H*. Lines *FI* and *EJ* are the straight lines parallel with Line *GH* and intersect with Lines *BG* and *AH* at Points *P*, *I*, *J* and *O*, respectively. *K* is the midpoint of straight line segment *AH*. Line *DR* passes through Points *D* and *K*. Line *EQ* is the straight line parallel with Line *DR* and intersects with the palm boundary at Point *Q*. Using these points and lines, we define the principal lines as below:

The **heart line** is a smooth curve that satisfies the following conditions (Figure 11-2 (a)):

i) Originating from Region *GHIP*;
ii) Running across Line-segment *OJ*;
iii) Not running across Line-segment *AH*.

The **head line** is a smooth curve that satisfies the following conditions (Figure 11-2 (b)):

(1) It is not the same curve as the extracted heart line;
(2) Originating from Region *ABML*;
(3) Running across Straight-line *DR*;
(4) The straight line which passes the two endpoints of this curve runs across Line-segment *EQ*;
(5) The straight line which passes the two endpoints of this curve does not run across Line-segment *BG*.

The **life line** is a smooth curve that satisfies the following conditions (Figure 11-2 (c)):

1) Originating from Region *ABML*;
2) Running across Line-segment *AH*;
3) The straight line which passes the two endpoints of this curve does not run across Line-segment *EQ*.

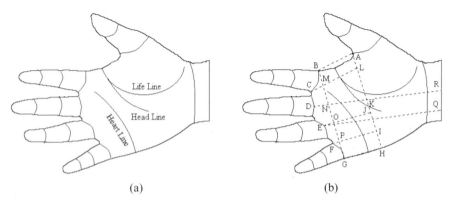

(a)   (b)

Figure 11-1. (a) The typical principal lines on a palm; (b) Defined points and lines on a palm.

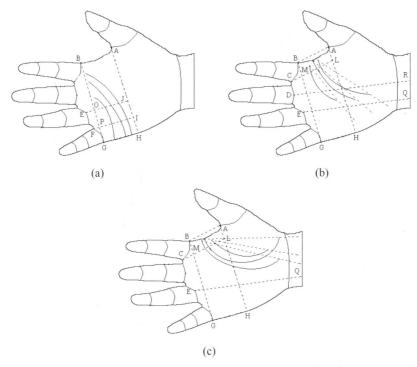

(a)   (b)

(c)

Figure 11-2. Definitions of principal lines: (a) the heart line, (b) the head line and (c) the life line.

Key Points Detection

Given the above definitions, we must first detect a set of points and lines before we can extract principal lines. Points *A*, *B*, *C*, *D*, *E*, *F* and *G* are the key points from which the other points and the defined lines can be easily obtained using their definitions (Figure 11-1 (b)). All images used in this chapter were obtained from the second and third versions of the online palmprint acquisition device, i.e. databases **PolyU-ONLINE-Palmprint-I** and **PolyU-ONLINE-Palmprint-II**. Although there are no pegs for the guidance of palmprint placement in the **PolyU-ONLINE-Palmprint-I** database, the enrolee has been told to stretch the fingers apart so that we can get enough information from the finger holes for the palmprint segmentation, i.e. key points detection. For the **PolyU-ONLINE-Palmprint-II**, there is a flat platen surface designed to guide the placement of a palm, so that the palmprints acquired are well aligned which is good for palmprint segmentation.

Detecting key points begins with extraction of the boundary of the palm by first smoothing the original image (Figure 11-3 (a)) using a low-pass filter and a threshold to convert it into a binary image (Figure 11-3 (b)) and then tracing the boundary of the palm (Figure 11-3 (c)). In our databases, a small portion of some palmprint images cannot obtain the area below the little finger. In these cases, we use the corresponding boundary segment of the image to represent the missing palm boundary segment.

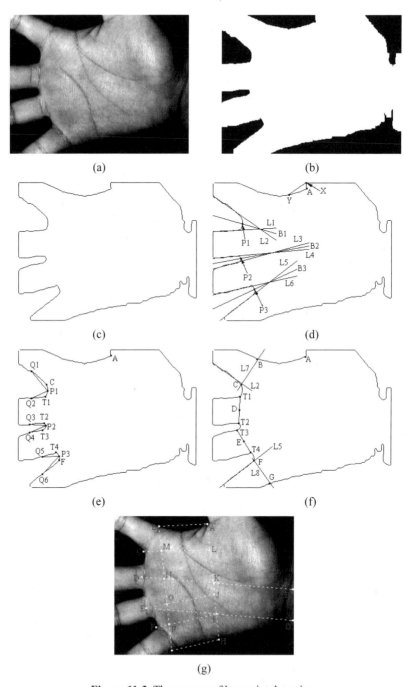

Figure 11-3. The process of key point detection.

In our work, we describe principal lines according to following three basic rules. First, the number of each type of principal line occurring in a palmprint is less than or equal to 1. If more than one line satisfies the same conditions, we keep the one with the greatest average magnitude. Second, we do not take into account the broken principal lines. When a principal line is broken at some places, we regard the broken point as its endpoint. Third, we regard each principal line as a curve without branches. Thus, if there are branches, we keep the smoothest curve and discard the others.

I. Detecting Point A (**Figure 11-3 (d)**)
1) Find a point (X) on the thumb boundary;
2) Find the point (Y) on the palm boundary whose column index is less l than that of Point X (here, $l = 30$);
3) In all points on the palm boundary segment between Point X and Point Y, find the point which is farthest from Line XY as Point A.

II. Detecting Points C, D, E and F
1) Use straight lines, L_1, L_2, L_3, L_4, L_5 and L_6, to fit the segments of the boundary of the index finger, middle finger, ring finger and little finger (Figure 11-3 (d)):

$$L_i: \quad y = k_i x + b_i, \tag{11-1}$$

$$k_i = \frac{\sum_{k=1}^{M_i} x_i^k \times \sum_{k=1}^{M_i} y_i^k - M_i \times \sum_{k=1}^{M_i} (x_i^k \times y_i^k)}{\left(\sum_{k=1}^{M_i} x_i^k\right)^2 - M_i \times \sum_{k=1}^{M_i} (x_i^k)^2}, \tag{11-2}$$

$$b_i = \frac{\sum_{k=1}^{M_i} y_i^k - k_i \times \sum_{k=1}^{M_i} x_i^k}{M_i}, \tag{11-3}$$

where $\{(x_i^k, y_i^k)\}_{k=1}^{M_i}$, $(i = 1, \cdots, 6)$ are the coordinates of the points on the segments of the boundary of the index finger, middle finger, ring finger and little finger, respectively; M_i is the total number of points on the corresponding segment.

2) Compute the bisectors (B_1, B_2 and B_3) of the angles formed by L_1 and L_2, L_3 and L_4, and L_5 and L_6:

$$B_i: \quad y = K_i x + B_i, \tag{11-4}$$

$$K_i = \frac{k_{2\times i-1} \times \sqrt{1 + k_{2\times i}^2} + k_{2\times i} \times \sqrt{1 + k_{2\times i-1}^2}}{\sqrt{1 + k_{2\times i-1}^2} + \sqrt{1 + k_{2\times i}^2}}, \tag{11-5}$$

$$B_i = \frac{b_{2\times i-1} \times \sqrt{1+k_{2\times i}^2} + b_{2\times i} \times \sqrt{1+k_{2\times i-1}^2}}{\sqrt{1+k_{2\times i-1}^2} + \sqrt{1+k_{2\times i}^2}}, \qquad (11\text{-}6)$$

where $i = 1,2,3$; $k_1 - k_6, b_1 - b_6$ are computed by Equations 11-2 and 11-3. The intersections of B_1, B_2 and B_3 with palm boundary segments between the fingers are Points P_1, P_2 and P_3, respectively (Figure 11-3 (d));

3) Find the points Q_1 and Q_2 on the boundary of index finger and middle finger, whose column index is less l_1 than that of Point P_1 (here $l_1 = 30$), and line up P_1Q_1 and P_1Q_2 (Figure 11-3 (e));

4) In all points on the finger boundary segment between Points P_1 and Q_1, find the point which is farthest from Line P_1Q_1 as one root point of the index finger, C (Figure 11-3 (e));

5) In all points on the finger boundary segment between Points P_1 and Q_2, find the point which is farthest from Line P_1Q_2 as one root point of the middle finger, T_1 (Figure 11-3 (e));

6) The root points of the middle, ring and little fingers, T_2, T_3, T_4 and F, are obtained using the same technique as described in Steps 3-5 (Figure 11-3 (e));

7) Line up T_1T_2 and T_3T_4, and take their midpoints as Points D and E (Figure 11-3 (f));

III. Detecting Points B and G (Figure 11-3 (f))

(1) Draw the perpendicular of Line L_2 from Point C, L_7, and intersect with palm boundary at Point B;

(2) Draw the perpendicular of Line L_5 from Point G, L_8, and intersect with palm boundary at Point G.

Figure 11-3 (g) shows the palmprint overlaid with the line definitions and the respective key points obtained.

11.2   Principal Lines Extraction

Palm lines, including the principal lines and wrinkles, are a kind of roof edge. A roof edge is generally defined as a discontinuity in the first-order derivative of a gray-level profile [68]. In other words, the positions of roof edge points are the zero-cross points of their first-order derivatives. Moreover, the magnitude of the edge points' second-derivative can reflect the strength of these edge points [69]. We can use these properties to design principal lines detectors. Given that the directions of principal lines are not constant, we should design line detectors in different directions and then apply a suitable detector according to the local information of the principal lines. Details of the principal lines extraction and representation are reported in Chapter 8 Section 8.1.

Extracting Potential Line Initials of Principal Lines

The heart line is defined as a curve originating from Region *GHIP* (Figure 11-2 (a)) and the life line and head line are defined as the curves originating from Region

ABML (Figure 11-2 (b) and (c)). Therefore we can extract the beginnings ("line initials") of the principal lines from these regions and then use these initials as the basis to extract the complete principal lines. A careful examination of a palmprint reveals that each principal line will initially run almost perpendicular to its neighboring palm boundary segment – approximated here by Line *AB* (life and head line) or Line *GH* (heart line) (Figure 11-3 (g)). If we denote the slope angle of the corresponding line (Line *AB* or Line *GH*) as θ, then the directions of the line initials of the principal lines are close to $\theta + 90°$, so we first extract all lines in this region using the $\theta + 90°$ line detectors $H_1^{\theta+90°}$, $H_2^{\theta+90°}$ (details of these line detectors can be found in Chapter 8, Section 8.1). Each of these extracted line segments is a potential line initial of principal lines. Hence, we should extract lines from each of these line initials and then keep the principal lines according to their definitions. Figure 11-4 shows these extracted lines: (a) is the original palmprint and (b) is the palmprint overlaid with the extracted potential line initials of the principal lines.

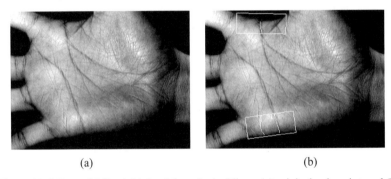

(a)   (b)

Figure 11-4. Potential line initials of the principal lines: (a) original palmprint, and (b) palmprint overlaid with the extracted potential line initials of the principal lines.

Extracting Heart Lines

Because principal lines do not curve greatly, it is a simple matter to use the current extracted part of the line to predict the position and direction of the next part on line tracing. Now, based on the extracted potential line initials, we designed a recursive process to extract the whole heart line, part by part.

Suppose that Curve *ab* in Figure 11-5 (a) is the extracted part of the heart line of the palmprint shown in Figure 11-4 (a). To extract the next part of the heart line, we trace back the extracted heart line *ab* from Point *b* and get the Kth point *c* (here *K* = 20). Since the heart line does not curve greatly, the region of interest (ROI), in which the next segment of heart line would be located, can be defined as a rectangular region *L* × *W* whose center point is Point *b*. Point *c* is the midpoint of one border whose length is *W*. Notice that *W* is a predefined value (here *W* = 20), and *L* equals to twice the distance between Points *b* and *c* (Figure 11-5 (b)).

Joining Points *b* and *c* gives us a straight line *cb*. The slope angle of straight line *cb* is α. Because principal lines curve so little, the direction of the next line segment should not vary much. Therefore we employ directional line detectors H_1^α, H_2^α

(details of these line detectors can be found in Chapter 8, Section 8.1) to extract the line segments in this ROI and then keep all of the branches connecting with *ac* (Figure 11-5 (c)). If only one branch connected with *ac*, this branch is regarded as the next line segment. Otherwise, we choose one branch as follows.

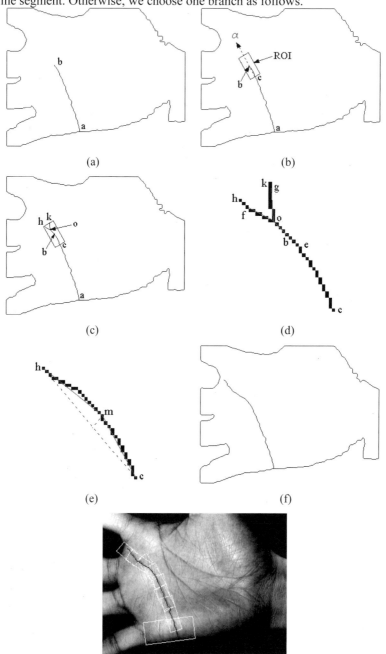

(a)   (b)

(c)   (d)

(e)   (f)

(g)

Figure 11-5. Heart line extraction.

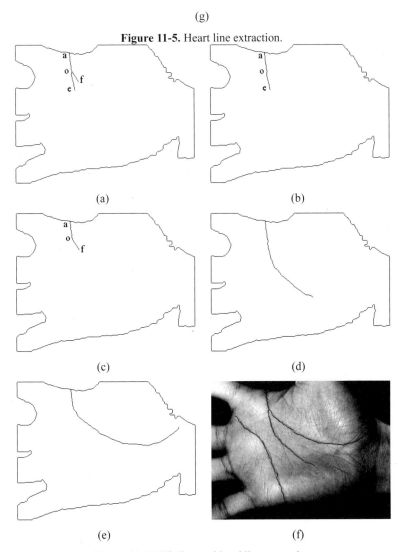

(a)

(b)

(c)

(d)

(e)

(f)

Figure 11-6. Life line and head line extraction.

In Figure 11-5 (c), two branches, *cok* and *coh*, are connected with *ac* in ROI, where Point *o* is the branch point. Figure 11-5 (d) shows the enlarged version of the ROI. We trace the line *oh*, *ok* and *oc* from Point *o* and get the Nth points *f*, *g* and *e*, respectively (here $N = 10$), and then line up *of*, *og* and *oe*, and compute Angle *foe* and Angle *goe*. The branch (*oh*) corresponding to the maximum angle (*foe*) is chosen as the next line segment for the tracing.

After obtaining the next line segment, we should determine whether the heart line reaches its endpoint. We regard the heart line as having reached its endpoint if the line *ch* in the ROI satisfies one of the following two conditions:

1) If the minimum distance from endpoint *h* to three sides of the ROI (not including the side passing through Point *c*) exceeds a threshold T_d (here $T_d = 5$), Point *h* is the endpoint.

2) If Angle *cmh* is less than a threshold T_a (here $T_a = 135°$), Point *m* is the endpoint (Figure 11-5 (e)). Here, point *m* on curve *ch* is the farthest point to the straight line ch.

If neither of these conditions are satisfied, we take the longer curve *ah* as the current extracted heart line and repeat this process recursively until the extracted curve reaches its endpoint. Figure 11-5 (f) shows the whole extracted heart line and (g) is the palmprint overlaid with the whole heart line and all of the ROIs involved in this heart line extraction.

Extracting Life and Head Lines
The process of extracting the life line and the head line differs little from that of extracting the heart line. The difference is in the rule that only one line may be extracted from each line initial. While this works well in heart line extraction, it is unsuitable for life line and head line extraction because life line and head line may share their line initials. Given this, we apply our observation that the branch point of the life line and head line should not exceed Line-segment *NK* and *LK* (Figure 11-1 (b)). With this in mind, when using the recursive heart line extraction process to extract the life line and head line, if the extracted curve does not run across Line-segment *NK* and *LK* and if there exists more than one branch, instead of choosing just one of the branches, we extract and trace the curves from each branch. If the extracted curve crosses Line-segment *NK* or *LK*, the extraction process is the same as for heart line extraction.

Figure 11-6 illustrates the process of life line and head line extraction. In this figure, (a) is the extracted line including two branches. In the original heart line extraction process, only one branch (*oe*) would be chosen and the other one (*of*) would be discarded. Obviously, it is wrong to discard branch *of* because it is a part of the life line. Since the extracted line does not run across Line-segment *NK* and *LK*, we split this branched curve into two curves *aoe* and *aof* (Figures 11-6 (b) and (c)) and extract the lines from each of them. In this figure, the extracted curve started with *aoe* is the head line (Figure 11-6 (d)) and the one started with *aof* is the life line (see Figure 11-6 (e)). Figure 11-6 (f) shows the palmprint overlaid with all of the extracted principal lines.

11.3   Six Types of Palmprints

The complete classification process for an input palmprint is as follows: 1) binary this palmprint and extract its boundary; 2) detect the key points; 3) extract heart line; 4) extract head line and life line; 5) calculate the number of principal lines and their intersections; and 6) classify the palmprint into one of the defined categories using the classification rules. We have already discussed the first four steps to extract the principal lines and to detect the key points. Now, we can classify palmprints by the number of the principal lines and their intersections (if any). Each type of principal line is less than or equal to 1, and there are at most three principal lines. Two principal lines are said to intersect only if some of their points overlap or some points of one line are the neighbors of some points of another line. If any two principal lines intersect, the number of intersections increases by 1. Therefore, the number of intersections of three principal lines is less than or equal to 3.

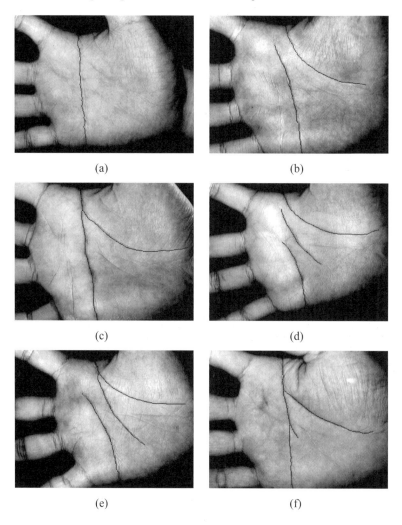

(a)

(b)

(c)

(d)

(e)

(f)

Figure 11-7. Examples of each palmprint category: (a) Category 1, (b) Category 2, (c) Category 3, (d) Category 4, (e) Category 5, and (f) Category 6.

Table 11-1. Palmprint classification rules.

No. of principal lines	≤ 1	2		3		
No. of the intersections of principal lines	0	0	1	0	1	≥ 2

Category	1	2	3	4	5	6

According to these definitions and findings, palmprints can be classified into the following six categories:

Category 1: Palmprints composed of no more than one principal line, as shown in Figure 11-7 (a);

Category 2: Palmprints composed of two principal lines and no intersection, as shown in Figure 11-7 (b);

Category 3: Palmprints composed of two principal lines and one intersection, as shown in Figure 11-7 (c);

Category 4: Palmprints composed of three principal lines and no intersection, as shown in Figure 11-7 (d);

Category 5: Palmprints composed of three principal lines and one intersection, as shown in Figure 11-7 (e);

Category 6: Palmprints composed of three principal lines and more than one intersection, as shown in Figure 11-7 (f).

Table 11-1 summarized these categories by the classification rules defined above.

11.4 Experimental Results

Our palmprint classification algorithm was tested on a database containing 13,800 palmprints captured from 1,380 different palms using databases **PolyU-ONLINE-Palmprint-I** and **PolyU-ONLINE-Palmprint-II**. In the first database, the images are 320 × 240 at 75 dpi, with eight bits per pixel. There are two different sizes of palmprints from the second database, 384 × 284 and 768 × 568. The palmprints with size 768 × 568 are resized to 384 × 284, so that all the palmprints are at 75 dpi with the same sizes. Then, the palmprints are classified and labeled manually into six Categories, in which 0.36% samples belong to Category 1, 1.23% to Category 2, 2.83% to Category 3, 11.81% to Category 4, 78.12% to Category 5 and 5.65% to Category 6. The distribution of each category in this palmprint database is listed in Table 11-2.

Correct classification takes place when the palmprint is classified into a category whose label is same as the label of this palmprint. Misclassification takes place when the palmprint is classified into a category whose label is different from the label of this palmprint. In all of the 13,800 palmprints in the database, 548 samples were misclassified: 7 in Category 1, 11 in Category 2, 25 in Category 3, 104 in Category 4, 349 in Category 5 and 47 in Category 6. The classification accuracy is about 96.03%. The confusion matrix is given in Table 11-3 and the classification accuracy is given in Table 11-4.

Table 11-2. Distribution of each category in our database.

Category	1	2	3	4	5	6
Number of palmprints	50	170	390	1,630	10,780	780

Percent (%)	0.36	1.23	2.83	11.81	78.12	5.65

Table 11-3. Classification results of the proposed algorithm.

Assigned category	Actual category					
	1	2	3	4	5	6
1	**43**	6	3	23	41	2
2	3	**159**	1	41	132	7
3	4	0	**365**	18	95	13
4	0	3	2	**1,526**	12	0
5	0	2	11	13	**10,431**	25
6	0	0	8	9	69	**733**

Table 11-4. Classification accuracy of the proposed algorithm.

Total samples	13,800
Correctly classified samples	13,252
Misclassified samples	548
Classification accuracy	96.03%

11.5 Summary

As the first attempt to classify low resolution online palmprints, this chapter presents a novel algorithm for palmprint classification using principal lines. Principal lines are defined and characterized by their positions and thickness. A set of directional line detectors is devised for principal line extraction. By using these detectors, the potential line initials of the principal lines are obtained. Based on these initials, the principal lines can be extracted completely using a recursive process. The local information about the extracted part of the principal line is used to decide a ROI and then a suitable line detector is chosen to extract the next part of the principal line in this ROI. After extracting the principal lines, we present some rules for palmprint classification. The palmprints are classified into six categories according to the number of the principal lines and their intersections. In our database, there are 13,800 palmprints, the distributions of Categories 1-6 are: 0.36%, 1.23%, 2.83%, 11.81%, 78.12% and 5.65%, respectively. The proposed algorithm classified these palmprints with 96.03% accuracy.

12 HIERARCHICAL PALMPRINT CODING WITH MULTI-FEATURES

In this chapter, we report a hierarchical multi-feature coding scheme to facilitate coarse-to-fine matching for efficient and effective palmprint verification and identification in a large database. Four levels of features are defined: global geometry based key point distance (Level 1 feature), global texture energy (Level 2 feature), fuzzy "interest" line (Level 3 feature) and local directional texture energy (Level 4 feature) [140]. In contrast with existing methods [31, 50, 65, 80, 119, 134-136] which employ a fixed mechanism for feature extraction and similarity measurement, we extract multiple features and adopt different matching criteria at different levels to achieve high performance using a coarse-to-fine guided search. Section 12.1 introduces this approach. Section 12.2 reports the hierarchical palmprint coding scheme which embraces global geometry boundary segments, global texture energy, fuzzy "interest" lines and local directional texture energy. Section 12.3 describes the use of multi-level features for matching measurement. Section 12.4 summarizes a guided searching strategy for the best match. Section 12.5 reports the experimental results and Section 12.6 summarizes this chapter.

12.1 Introduction

At a first glance, palmprint and fingerprint patterns appear to resemble each other in some ways. Both consist of a large amount of ridges. Although the minutiae based matching which utilizes terminations and bifurcations of the ridges is powerful for fingerprint verification and identification, such an approach is not suitable for palmprint patterns due to the change of orientations. Zhang and Shu [18] proposed to use datum point invariance and line feature matching for palmprint feature extraction and verification. They introduced a directional projection algorithm to localize the datum points for palmprint registration, then, a set of fixed masks is used to detect feature lines for matching. However, there is a problem when the principal lines are not clear which may affect their results.

Recently Duta *et al.* investigate the feasibility of matching palmprints based on feature points [77]. Instead of extracting feature lines explicitly as in [18], they apply the following six steps to extract isolated feature points that lie along palm lines: (a) palm image smoothing; (b) image binarization by interactively thresholding the smoothed grey palm image; (c) successive morphological erosions, dilations and subtractions for feature point detection; (d) location adjustment for the feature points; (e) calculation of the orientation of each feature point (the orientation of the palm line which a feature point is associated with); and (f) removal of redundant points. They use the point matching technique to determine whether two sets of feature points/orientations are the same by computing a matching score. Although this approach has the advantages of simplicity and reasonable accuracy, it lacks robustness in the following aspects: (i) the detection of feature points rely on the selection of threshold value for image binarization and the consequent successive morphological operations; (ii) the feature point matching is based on the traditional exhaustive comparison method, which is very time consuming and may not meet the real-time requirement for on-line matching of a large collection of palmprint patterns.

Chapter 5 Section 5.3 already reported a feature point called interesting point detected by Plessey operator, which is better than the edge points to represent palmprint images for matching. However, it is a time consuming process for matching on these feature points, which is not suitable for searching a palmprint in a large database. It is known that dynamic feature extraction, guided search and knowledge-based hierarchical matching will have significant potential in enabling both image identification and image classification to be performed more effectively [82]. However, most of the existing systems and research adopt a fixed scheme for feature extraction and similarity measurement, which are not suitable to search a palmprint image in a large database. Our previous work [82] has explored these issues (i.e. guided search by multiple features), however, it is only working on the offline palmprints. We here to propose a hierarchical palmprint coding scheme to facilitate coarse-to-fine matching [140], focused on the online palmprint images, for efficient and effective identifying a palmprint in a large database. More specifically, we extract different palmprint features at different levels: Level 1 feature: global geometry based key point distance; Level 2 feature: global texture energy; Level 3 feature: fuzzy "interest" line; and Level 4 feature: local directional texture energy vector for fine palmprint matching. We start with global geometry feature to localize the region of interest of palmprint sample at coarse level and apply a distance measurement of palm boundary to guide the dynamic selection of a small set of similar candidates from the database for further processing. We also use the global texture energy (GTE) for fast search for the best match. Such a mask-based texture feature representation is characterized with high convergence of inner-palm similarities and good dispersion of inter-palm discrimination. We then adopt fuzzy set theory to detect "interest" feature lines to guide the search for the best match at fine level. Finally, we apply local texture measurement to establish a feature vector for palmprint matching.

This hierarchical palmprint identification system consists of four components, palmprint acquisition, preprocessing, feature extraction and hierarchical matching. In our CCD based palmprint capture device [65], there are some pegs between fingers to guide the palm's stretching, translation and rotation. These pegs separate the fingers, forming holes between the index finger and the middle finger, and between the ring finger and the little finger. Different preprocessing approaches have been proposed including key point-based and datum point based approaches [18, 65-66, 82]. In this chapter, we use the preprocessing technique described in Chapter 6 Section 6.3 to align the palmprints [50]. In this technique, the tangent of two finger holes are computed and used to align the palmprint. The central part palmprint sub-image, 128 × 128, is obtained to represent the whole palmprint. Such preprocessing greatly reduces the translation and rotation of the palmprints captured from the same palms. Feature extraction is to obtain multi-level features for palmprint representation.

Hierarchical matching is to determine the similarity of two palmprint images for identity authentication. Figure 12-1 illustrates the general structure of our system. Initially, a palmprint image is captured by our palmprint scanner [65]. Then, the boundaries between fingers are extracted. Based on the boundaries, the two key points can be determined so as to setup the coordinate system to extract the central parts. The distance of two key points is considered as Level 1 feature. Based on the central part palmprint sub-images, global texture energy, fuzzy "interest" line and local directional texture energy are also extracted. The features in the hierarchical feature database are retrieved and compared with input features, in a multi-level fashion.

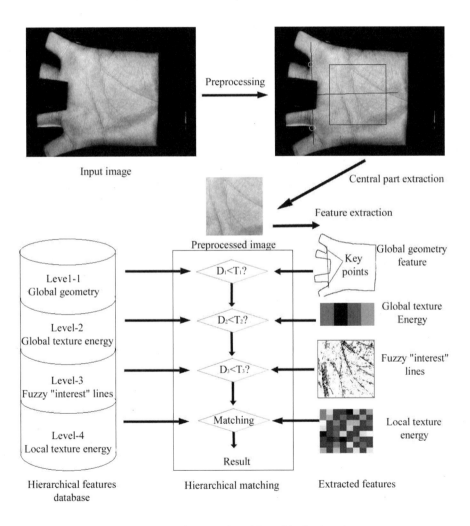

Figure 12-1. System diagram of our hierarchical palmprint system.

12.2   Hierarchical Palmprint Coding

Feature extraction is a key issue for pattern recognition and palmprint consists of very complicated patterns. It is very difficult, if not impossible, to use one feature model for palmprint matching with high performance in terms of accuracy, efficiency and robustness. Although the research reported in [77] resulted in a more reliable approach to palmprint feature point detection than the line matching algorithm detailed in [18], the issues of efficiency and robustness remain untackled. The technique reported in [77] involves one-to-one feature point based image matching, which requires high computation resource for a large palmprint database. In addition, the matching lacks flexibility because only one similarity measurement is applied. In order to speed up the searching process for the best match with reliable features and flexible matching criteria, we adopt a multi-level feature extraction strategy and a flexible similarity measurement scheme. Instead of using a fixed feature extraction mechanism and a single matching criterion as in [18, 77], we use a hierarchical coding scheme to extract multiple palmprint features at both global and local levels, in which levels 1 and 2 belong to global features while levels 3 and 4 belong to local features. Detail design of these four levels of palmprint feature extraction strategies are shown as follow.

Level 1: Global Geometry Feature – Key Point Distance
The key point distance is measured based on boundary segments between fingers. To obtain a stable palmprint image for reliable feature extraction, some pegs are designed on the user interface of the flat platen surface of our proprietary design palmprint acquisition system. These pegs can control the placement of the user's palm. In addition, a coordinate system is defined to align different palmprint images for feature measurement. Figure 12-1 illustrates the global geometry feature of fingers. Such boundary segments can be obtained by a boundary tracking algorithm with the following major steps:

Step 1: Convert an original palmprint image to a binary image by convolving a
            lowpass filter with the original image and thresholding.
Step 2: Apply boundary tracing to obtain the boundaries between fingers.
Step 3: Compute the tangent of line segments.
Step 4: Identify the two key points and calculate their distance as the global
            geometry feature.

Level 2: Global Texture Feature – "Tuned" Mask Based Texture Energy
For the purpose of fast selection of a small set of similar palmprint patterns from the database, previously we applied four "tuned" masks to capture the global palmprint texture features which are more sensitive to horizontal lines, vertical lines, 45^0 lines and -45^0 lines respectively [45, 139, 141]. In this approach, the local variance after convolution is well-approximated by the sum of squared values of convolved image within the test window, which is expressed as:

$$TE(i,j) = \frac{\sum_{W_x}\sum_{W_y}(I * A_k)_{rs}^2}{P^2 W_x W_y},$$ (12-1)

where the *rs* sum is over all pixels within a window W of size W_x x W_y centered on the pixel at i, j; A_k is a zero sum "tuned" 5×5 convolution mask and P is the parameter normalizer, $P^2 = \sum_{i,j}(A_{i,j})^2$. Such a texture energy measurement for global palmprint feature extraction has the following characteristics: (a) insensitive to noise; (b) insensitive to shift changes; (c) easy to compute; and (d) high convergence within the group and good dispersion between groups. In view of those advantages stated, we defined 4 "tuned" masks as shown in Figure 12-2 to extract the global texture feature for Level 2 feature representation. The horizontal global texture energy is the mean all the texture energy (TE) obtained from the given horizontal "tuned" masks, similarly for the vertical, $45°$ and $-45°$ global texture energy. The four directional global texture energies constitute the global texture energy.

-1	-2	-4	-2	-1		-1	0	2	0	-1
0	0	0	0	0		-2	0	4	0	-2
2	4	8	4	2		-4	0	8	0	-4
0	0	0	0	0		-2	0	4	0	-2
-1	-2	-4	-2	-1		-1	0	2	0	-1
		(a)						(b)		
0	-1	-4	0	2		2	0	-4	-1	0
-1	-6	0	8	0		0	8	0	-6	-1
-4	0	12	0	-4		-4	0	12	0	-4
0	8	0	-6	-1		-1	-6	0	8	0
2	0	-4	-1	0		0	-1	-4	0	2
		(c)						(d)		

Figure 12-2. Four kinds of "tuned masks" for global palmprint texture extraction. (a) Horizontal line, (b) Vertical line, (c) $45°$ line, and (d) $-45°$ line.

Level 3: Fuzzy "Interest" Lines
The so-called "interest" lines refer to the dominant feature lines such as principal lines, wrinkles, etc. in palmprint. The major steps of extracting dominant lines are listed below.

Step 1   Convert the pre-processed images into feature image as described in [142].
Step 2   Apply the fuzzy rule to extract the "interest" lines. Let I_T be a feature image. The fuzzy output is given by the following piecewise linear membership function:

$$S(l) = \begin{cases} 0 & \text{if} & l < a \\ (l-a)/(b-a) & \text{if} & a \le l \le b \\ 1 & \text{if} & b > 1 \end{cases} \quad (12\text{-}2)$$

The parameters, a and b are depending on the feature images. We set the

$a=\mu$ and $b=\mu+\sigma$, where μ and σ are the sample mean and standard deviation of $I_T(x,y)$, respectively. A sample of fuzzy interest lines is shown in Figure 12-3.

Step 3   Take the mean of the fuzzy output in a small window to represent the local interest lines such as

$$v(I_T) = \frac{1}{MN}\sum_x\sum_y S(I_T(x,y)), \qquad (12\text{-}3)$$

where the size of the local block is M by N. In our experiments, both M and N are set to 20. Finally, the "interest" lines can be represented by a 64 dimension feature vector obtained from 64 overlapped blocks.

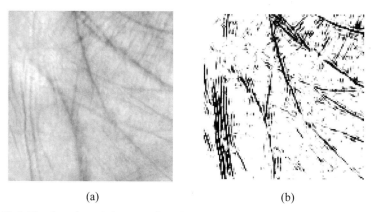

(a)   (b)

Figure 12-3. The detection of "interest" lines based on fuzzy theory. (a) Original image, and (b) its "Interest line".

Level 4: Local Directional Texture Energy

This level is the last level to make a final decision so that high accuracy of this level is expected. To achieve this goal, we generate a long feature vector from local directional texture energy so as to facilitate more information. The local directional texture energy can be obtained by:

$$u_i = \frac{1}{XY}\sum_x\sum_y\left|\sum_{W_X}\sum_{W_Y}(I * A_i)_{rs}\right|, \qquad (12\text{-}4)$$

where X by Y is the size of the local block for computing the local directional texture energy. We set $X = 20$ and $Y = 20$, which balances the size of feature vector and accuracy of the system. Therefore, each block is overlapped with its adjacent blocks.

12.3   Multiple Similarity Measurement

To avoid the blind search for the best fit between the template pattern and all of the sample patterns stored in an image database, a guided search strategy is essential to reduce computation burden. We propose here to consider multiple palmprint features and adopt different similarity measures in a hierarchical manner to facilitate a coarse-to-fine palmprint matching scheme for personal identification. As stated in previous section, four palmprint features are extracted, Level 1: global geometry feature, Level 2: global texture energy, Level 3: "interest" lines, and Level 4: local directional texture energy. Levels 1 and 2 used to identify a small set of the most relevant candidates at coarse level. The local line based matching and feature vector comparison from Levels 3 and 4 are performed at fine level. Unlike the matching scheme described in [18], our fine matching is conducted in a hierarchical manner based on different features. Detail descriptions on these matching schemes are listed as follow.

A. Minimum Distance for Global Feature Matching

We have defined two global features, 1) geometry distance of key points, and 2) global texture energy. Our goal is to search a palmprint in a large database; as a result, the searching speed is one of the main concerns. Simple and effective distance measures are expected. The distance measure for the geometry feature vectors is:

$$D_1 = \left| d_i - d_j \right|, \tag{12-5}$$

where i and j represent two palmprint images and d_i (d_j) is the distance between key points of the i (j) palmprint images.

The global texture energy can be measured by distance similarity measurement. Let the global texture energy vector of ith palmprint image be $[v_{0i}, v_{1i}, v_{2i}, v_{3i}]$, similarly for j^{th} palmprint image. Their similarity can be measured by,

$$D_2 = \sum_{k=0}^{3} \left| v_{ki} - v_{kj} \right|. \tag{12-6}$$

B. Minimum Distance for Local Feature Matching

In addition to the two global features, we also designed two local features in our coding scheme, "interest" lines and local directional texture energy. The "interest" lines are represented by a 64 dimensions feature vector. At this level, we use an angular distance to evaluate the difference between two feature vectors. Let X and Y be two "interest" lines. The angular distance is defined as

$$D_4 = \frac{X^T Y}{\|X\| \|Y\|}. \tag{12-7}$$

In the local directional texture energy, we use a local angular distance to evaluate the difference between two feature vectors. For simplicity, let the local directional texture energy vector of ith palmprint image be $[y_{0i}, y_{1i}, y_{2i}, y_{3i} \ldots y_{64i}]$, where $y_{ki}=[u_0, u_1, u_2, u_3]$ where u_k is defined in Equation 12-4. The local angular distance is defined as:

$$D_4 = \frac{1}{64} \sum_{k=1}^{64} \frac{y_{ki} y_{kj}^T}{\|y_{ki}\| \|y_{kj}\|}. \tag{12-8}$$

12.4 Hierarchical Palmprint Matching

Coarse Level Similar Palmprint Pattern Classification

This classification task can be viewed as a decision making process that locates an input palmprint sample to those palmprint images with the similar measurements in the database. It is very important to select suitable features that can discriminate different palmprint categories. Therefore, the choice of the palmprint features must be as compact as possible, and yet as discriminating as possible [143]. In other words, the feature measurements of palmprint samples with distinct texture primitives should exhibit large variances while the measurements of the similar patterns should possess very small diversity. Thus, such a global feature is characterized with high convergence of inner-palm similarities and good dispersion of inter-palm discrimination. Figure 12-4 shows three palmprint samples from the different individuals with distinctive texture features and Figure 12-5 demonstrates the distribution of global palmprint texture energy measurements. Some palmprints may explicit similar palmprint patterns. To tackle such a problem, we propose a dynamic selection scheme to obtain a small set of the most similar candidates in the database for further identification by image matching at fine level. Initially, we search for the best similar palmprint matching sub-set with Level 1 global geometry feature. Our similarity measurement method is based on the comparison of key point distance with respect to its length of different samples. The candidates with small distance difference such as $D_1 < T_1$, will be considered for further coarse-level selection by global texture energy. The idea behind this is to eliminate those candidates with large difference of Level 1 global geometry feature and generate a list of the very similar candidates with very small difference of their key point distance. The candidates in the list will undergo further selection in terms of its Level 2 global texture energy (GTE). Only those samples that remain very close GTEs will be considered for fine-level matching.

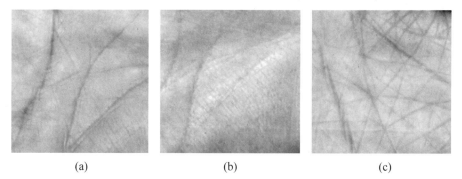

(a) (b) (c)

Figure 12-4. Samples of different palmprint patterns with distinctive texture features. (a) strong principal lines, (b) less wrinkles and (c) strong wrinkles.

Fine Level Multi-step Image Matching

The proposed fine matching algorithm starts with simple distance measure for "interest" line. If the matching score D_3 is larger than the threshold T_3, controlling the false acceptance and false rejection rates at this level, the palmprint images will go through the final fine-matching in terms of their local directional texture energy. This matching is based on the comparison of local feature vectors in terms of their local angular distance. The best match is the candidate with the least distance.

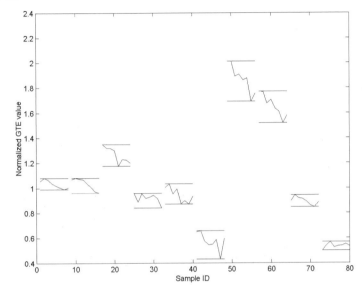

Figure 12-5. Comparison of palmprint GTE distribution: inter-palm dispersion vs. inner-palm convergence. This figure shows the distribution of GTE from 80 palmprint images from 10 palms.

12.5   Experimental Results

A large palmprint database is formed by taking 5,437 palmprint samples from the **PolyU-ONLINE-Palmprint-II**, for the palmprint matching and verification tests. Samples of such palmprint images are shown in Figure 12-4. A series of experiments have been carried out to verify the high performance of the proposed algorithms using this database. The dynamic features selection is demonstrated by a multi-level palmprint feature extraction for personal identification and verification (see our work on palmprint verification [82]). The experiment is carried out in two stages. In stage one, the global palmprint features are extracted at coarse level and candidate samples are selected for further processing. In stage two, the regional palmprint features are detected and a hierarchical image matching is performed for the final retrieval.

The verification accuracies at different levels are shown in Figure 12-6. Figures 12-6 (a), (c), (e) and (g) present the probability distributions of genuine and impostor at different feature levels. The corresponding receiver operating characteristic (ROC) curves, being a plot of genuine acceptance rate against false acceptance rate for all possible operating points are demonstrated in Figures 12-6 (b), (d), (f), and (h). Based on the ROC curves, we conclude that Level 4 local texture is the most effective feature. The Level 3 fuzzy "interest" lines are better than Level 2 global texture energy. The Level 1 geometry information only provides limited classification power.

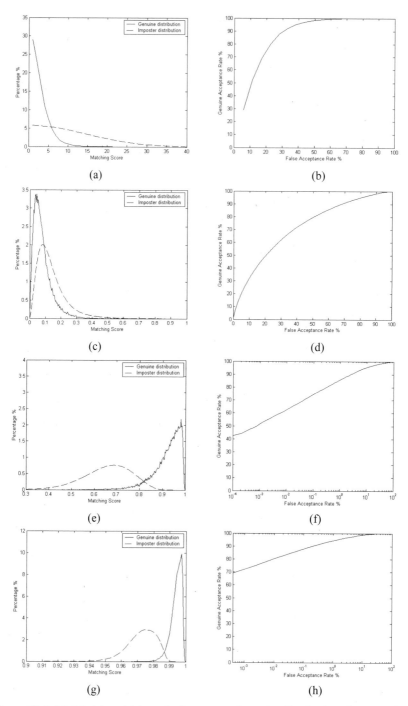

Figure 12-6. (a), (c), (e), and (g) are the imposter and genuine distributions from Level 1 to Level 4 features, respectively. (b),(d), (f), and (h) are their ROC curves.

To evaluate the verification accuracy of our hierarchical palmprint system, we use three parameters, T_1, T_2 and T_3, which control the false acceptance and false rejection rate of the first three levels of feature representation. Three sets of parameters, T_a, T_b and T_c are tested. Table 12-1 lists these parameters and the corresponding ROC curves are illustrated in Figure 12-7 for comparison. It is shown that Level 4 local texture feature performs better than our hierarchical palmprint approach when the false acceptance rate is large, such as 5%. However, a biometric system always operates in the condition of low false acceptance rate. Apparently, our hierarchical palmprint approach and fine-texture method have the similar performance when they keep low false acceptance rates. The false rejection and correct rejection rates of the first three levels are given in Table 12-2.

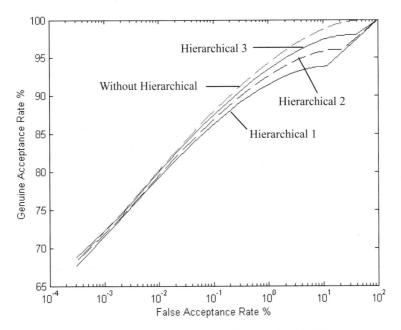

Figure 12-7. The ROC curves of our hierarchical approach with different parameters.

Table 12-1. The selection of parameters T_1, T_2, T_3 and T_a T_b, T_c.

	T_a	T_b	T_c
T_1	20	23	25
T_2	0.0194	0.0216	0.0250
T_3	0.764	0.73	0.68

The proposed approach is implemented using Visual C++ 6.0 on a personal computer with Intel Pentium III processor (500MHz). The execution time for the preprocessing, feature extraction and matchings are shown in Table 12-3. The total execution time is about 0.7 second, which is fast enough for real-time verification. Based on the computing time in Table 12-3, we can estimate the computing time for searching a palmprint in a large database. Let P_1, P_2 and P_3 be the percentages of the total number of palmprint samples in the database removed by Level 1, Level 2 and Level 3 features, respectively. They depend on the thresholds T_1, T_2 and T_3. Also, let the size of database be D and the computation time for preprocessing, feature extraction, Level 1 matching, Level 2 matching, Level 3 matching and Level 4 matching be S_p, S_f, S_1, S_2, S_3, and S_4, respectively. Based on these variables, we obtain the following formulas for searching a palmprint image in a database with D images by sequential and hierarchical approaches. Sequential approaches:

$$T_S = S_p + S_f + D \times S_4. \tag{12-9}$$

Hierarchical approaches:

$$T_H = S_p + S_f + D \times S_1 + D \times (1 - P_1) \times S_2 + D \times (1 - P_1 - P_2) \times S_3 + D \times (1 - P_1 - P_2 - P_3) \times S_4. \tag{12-10}$$

Table 12-2. System performance from the first three levels of parameters.

	False Rejection Rate	*Correct Rejection Rate*
T_a	6.13%	88.23%
T_b	3.89%	78.19%
T_c	1.98%	60.51%

Table 12-3. System execution time.

Operations	Times (ms)
Preprocessing	538
Feature Extraction	215
Level 1 matching	$4.7 * 10^{-5}$
Level 2 matching	$3.4 * 10^{-4}$
Level 3 matching	0.009
Level 4 matching	0.059

Based on this two equations and the computation time listed in Table 12-3, we can estimate computation time for searching a palmprint in a large database with D images by sequential and hierarchical approaches when P_1, P_2, and P_3 are known. We use T_a, T_b, and T_c to obtain their corresponding P_1, P_2, and P_3 for Figure 12-8, plotting the database size D against the computation time. According to Figure 12-8, the execution time of our hierarchical coding scheme which includes preprocessing, feature extraction and matching for a simulated large database with 10^5 palmprint samples is 2.8 s while the traditional sequential approach requires 6.7 s with the equal error rate of 4.5%. It is obvious that our hierarchical approach is much effective than sequential method when the database is large.

Comparing with the existing techniques for palmprint classification and identification, our approach integrates multiple palmprint features and adopts a flexible matching criterion for hierarchical matching. Table 12-4 summarizes the major features of our method and the other techniques, [18] and [77], with respect to their testing database size, feature selection strategies, matching criteria, searching scheme and accuracy. The experimental results presented above demonstrate the improvement and advantages of our approach. For future work, a dynamic indexing will be developed to handle the expansion of the palmprint database with robustness and flexibility.

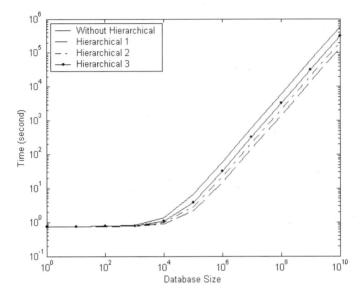

Figure 12-8. Computation time for large database.

Table 12-4. Comparison of different palmprint matching methods.

	Feature Point Based Matching [77]	Line Based Matching [18]	Hierarchical Matching (Proposed Approach)
Database Size	30 samples	200 samples	5,437
Feature Extraction	Feature points (single feature)	Lines (single feature)	Texture, geometry, lines (multiple features)
Matching Criteria	Distance measurement (fixed measurement)	Euclidian distance (fixed measurement)	Energy difference, angular difference (flexible measurement)
Search Method	One-to-one comparison (sequential)	One-to-one comparison (sequential)	Guided search (hierarchical)
Accuracy	good	limited	good

12.6   Summary

Palmprint is regarded as one of the most unique, reliable and stable personal characteristics and palmprint verification provides a powerful means to authenticate individuals for many security systems. Palmprint feature extraction and matching are two key issues in palmprint identification and verification. In contrast to the traditional techniques, we propose a hierarchical palmprint coding scheme to integrate multiple palmprint features for guided palmprint matching. The combination of four-level features including Level 1: global geometry feature; Level 2: global texture energy; Level 3: fuzzy "interest" line; and Level 4: local texture feature possesses a large variance between different classes while remaining high compactness within the class. The coarse-level classification by Level 1 and Level 2 features is effective and essential to reduce the number of samples significantly for further processing at fine level. Then, Level 3 and Level 4 local texture leads to fast search for the best match. The experimental results provide the basis for the further development of a fully automated palmprint-based security system with high performance in terms of effectiveness, accuracy, robustness and efficiency.

PART IV

PALMPRINT AUTHENTICATION SYSTEM

13 PALMPRINT ACQUISITION SYSTEM

Capturing high quality palmprint images in a very short period of time is a fundamental step in palmprint research. While it is very important to have a palmprint scanner of doing this in order to develop a real-time on-line palmprint identification system, limited research effort has been put into designing and implementing palmprint acquisition systems. Several companies (NEC and Identix) have developed inkless palmprint identification systems for criminal applications [111-112]. Their systems acquire high resolution images through live scanners, from which many detailed features, such as ridges, singular points and minutiae points, can be extracted. Several other research teams [65-67, 77, 134-135, 138] are working on recognition methods for inked and/or inkless palmprints. One of these teams [66] obtains inkless palmprint images using a digital scanner. However, the lengthy acquisition process and distorted palm skin that results from using a flatbed scanner suggest that it is not suitable for our intended purpose. The fact that there is no suitable palmprint acquisition system available on the market for civil applications has motivated us to propose this innovative palmprint acquisition system. In this chapter, we first go through the requirements analysis of a palmprint acquisition system. In Section 13.2 we report the parameter selection in the system design. In Section 13.3 we disclose an evaluation of the system performance. Section 13.4 summarizes the chapter.

13.1   System Requirements

As mentioned in Chapter 2, our online palmprint acquisition device has been evolved from different versions, with a great improvement on image quality and user interface. This section is going to analyze the design of the final version palmprint acquisition system, which is tailored for various civilian applications such as access control, ATM, etc. A typical flatbed scanner taking long time to obtain an image does not desire for our intended purpose. In addition, the distorted skin surfaces on the palmprint image degrade the quality of the palmprint features. Figure 13-1 shows a palmprint image obtained by a flatbed scanner in grayscale at a resolution of 150 dpi, which needs 10 seconds. In view of these circumstances, we need to design a novel palmprint acquisition system, which is different from the existing scanner based system. The objectives of the proposed system are:

1)   to acquire palmprint images in real time.
2)   to acquire palmprint images at high quality.
3)   to provide a balanced cost and quality, i.e. high quality with reasonable price.

4)   to provide a good user interface.

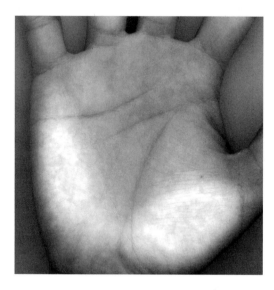

Figure 13-1. Palmprint image obtained by a scanner in grayscale level at 150 dpi resolution.

In order to prove the validity of our proposed system, a prototype was built. The innovative and distinctive features of the system include:

(1)   a specially designed user interface for the palmprint acquisition was proposed to minimize the image quality degradation due to the traditional glass-plate palm scanner.

(2)   the ability of instant palmprint image acquisition by the CCD camera at 25 frames per second, i.e. 40 ms per frame.

(3)   the ability of acquiring high quality palmprint features such as principal lines, wrinkles and ridge texture in small sized system at reasonable cost.

We need to pay special attention on the outlook design and the user interface design to gain higher user acceptance. Also, the mode of acquisition is important for both user and system, i.e., the user feels comfortable on using the device through the user interface, while the system can obtain distortion-free and noise-free images. In order to minimize the possible distortion due to the contact between the palm and the glass plate (traditional design), we use a new idea on the user interface so that the user's palm does not touch on any plate or surfaces. It avoids the latent prints formed on the glass also. Actually, the contactless design of the interface for palmprint acquisition plays an important role. Table 13-1 shows a set of requirements for the proposed system, with problems and solutions provided. The balanced system performance is the most important issue, as usually there are tradeoffs on different design, i.e., price vs. quality. Our target is to design the system with good balance of cost and quality, and to meet the requirements stated.

Table 13-1. Requirements analysis for the proposed palmprint acquisition system.

Requirements	Problems domain	Solutions
Uniform illumination	Type of light settings	Use ring shaped white fluorescent light
	Lighting scheme	Use ring light for sharpening the contrast of palmprint
Minimal distortion on the palm	Mode of acquisition	Contactless mode of acquisition
	Optical system design	Use larger lens format, optimal f-stop, etc
	User interface design	Use pegs to guide the users
Image quality	Optic's quality	Use higher quality of lens
	Optical system parameter selection	Careful selection on lens parameters such as lens format, f-stop, field of view, depth of field, etc.
	Minimum resolvable details	By defining the finest details to be acquired, and use the appropriate lens and CCD sensor.
	CCD sensor selection	Use 1/3" interline transfer black and white CCD camera, with 737×575 active pixels, where the pixel size is $6.6 \, \mu m \times 6.3 \, \mu m$.
Spatial resolution	Optical system design	The field of view, the sensor format, etc. The palm size to be imaged is $5.1" \times 3.8"$ while there are 768×568 pixels in the image plane after A/D, and the spatial resolution is 150 dpi.
Easy to use	User acceptability	A comfort panel for the user interface
	User interface design	An intuitive to use interface design
Effective palmprint data	User interface design	The mode of acquisition and the field of view from the optical system
Real time operation	Image size	Amount of information needed is limited to 768×576 pixels, which is enough for the palmprint feature, but not too large for computation requirement.
	A/D converter	Use a frame grabber at 25 fps (i.e. 40 ms) to acquire a palmprint image
	A/D conversion scheme	Use 8 bits per pixel (256 gray levels)
	Processing power	Processing power of the host uses a CPU of 950 MHz or higher, with 128 MB Ram.
Small in size	Optical system design	The system is designed as small as possible while it still keeps the image at high quality with minimal distortion. It is related to the lens focal length and type of lens selection.
Cost of the system	Components selection of the whole system	Choose components, at the best performance for price, off the shelf from the market to build the system.

System Framework
The schematic diagram of the proposed system is shown in Figure 13-2. There is a user interface for the input of the palm. The developed palmprint acquisition system mainly consists of a flat platen surface for the palm input, a light source unit, CCD camera, A/D converter, processing unit, memory, storage, keypad, LCD display, output signal and recognition engine. The ring light source is designed to provide white light source of high intensity, and it can increase the contrast of the palmprint features from the uneven palm skin surfaces. The optical system is fabricated in a controlled environment in order to minimize the effects of ambient light. When the signal is generated by the CCD, it is digitized by an 8-bit A/D converter and then the digitized data is transferred to the processor through the PCI bus. The palmprint image is stored in the storage of the system. An interface board is designed to communicate the user with the system through a keypad and a display unit. Recognition engine is used for the personal identification. The output signal is sent when a correct match is obtained. We have built a prototype of the system with the recognition engine design in [65] to provide access control to our laboratory. Details are shown in Chapter 14.

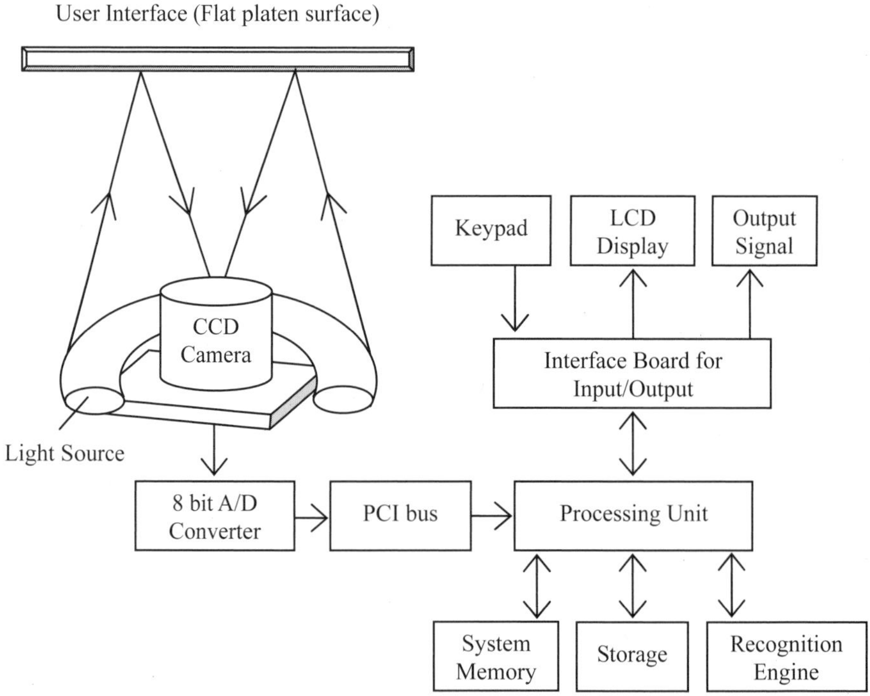

User Interface (Flat platen surface)

Figure 13-2. The schematic diagram of the palmprint acquisition system.

System Requirement and Analysis
Before we can go into the design stage, we have to answer to the following questions:
1. What is the minimum feature size to be acquired?

2. What is the minimum spatial resolution of the system for the required features?
3. How large is the palm size to be acquired?
4. What is the optical system parameters selection?
5. What kind of lighting is used for the illumination of a palm?

We can get the blueprint of the system by answering the above questions. The analysis starts with feature analysis so as to determine the finest feature to be captured. There are rich features in a palm, classified into three levels: principal lines, wrinkles and ridges. The first two levels of features are easy to obtain since they have a specific width and depth. For the ridge pattern, we found that even on the same palm, ridge-valley pattern at different part has different strength, i.e. some portion is very strong (with high contrast) while some portion is very blur (with low contrast). Most of the fingerprint sensors use 500 or more dpi for the image acquisition. Actually, the third level of feature appear on the palm (ridge) is the same type of skin of the fingerprint. It implies that if we would like to precisely obtain the ridge-valley patterns from a palm, we need about 400 dpi spatial resolutions. If the size of a full palm from an adult is 5" × 4" (estimated values), the image size becomes 2,000 × 1,600. For an 8 bits grayscale representation, it needs 3 MB for a palmprint image. By using this resolution, we need to employ a large format of sensor with higher price, and need to handle a file large in size with higher computation power. Apparently, there are tradeoffs among the resolution, computation and cost of the design. According to our objectives, we would like to take a lower resolution design with reasonable price, but still keep the rich palmprint features to achieve a balanced system.

13.2   Parameter Selection

Pixel Transition of Palmprint Features
According to the Nyquist-Shannon sampling theorem [113], when converting analog signal to digital signal, the sampling frequency must be greater than twice of the highest frequency of the input signal in order to be able to reconstruct the original signal from the sampled data. We first discuss the finest pattern – ridge-valley pattern. On average, one ridge and valley pair pattern in a fingerprint is about 1 mm width, according to [59]. We assume the ridge-valley pair on a palm is also spans 1 mm width. That is, if the system is able to reconstruct signals from a pair (1 cycle in 1 mm) of ridge-valley pattern from a palm, 2 sampling points (pixels) are needed on the sensor plane. We suppose that the ridge is white color and valley is black color so that 100% contrast can be obtained in this transition. Figure 13-3 shows the contrast analysis of the palmprint features. From the left of Figure 13-3 (a), the ridge-valley pattern exactly falls on the grid of the sensor array to provide the ideal contrast of the pattern. However, it is not always the case that they fall on the grid in the same phase, i.e. sometimes they fall on the sensor grid out-of-phase, as shown in the right of Figure 13-3 (a), so that the signal response is not discernable for the ridge-valley pattern. Figure 13-3(b) shows the principal line or wrinkle lies on the background. If it is out-of-phase, the line feature spans on two pixels, and the contrast is dropped to 50% of its original contrast.

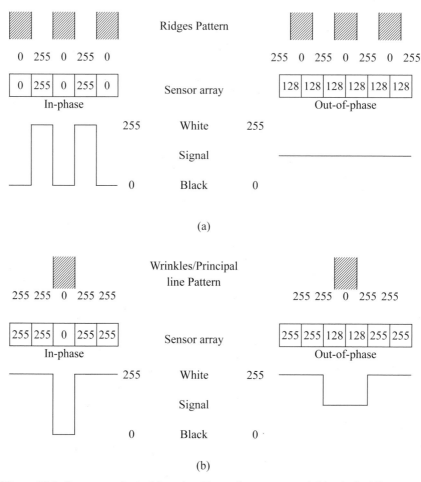

Figure 13-3. Contrast analysis, (a) on the ridge-valley pattern, and (b) principal line or wrinkle.

Our system uses a 1/3" interline CCD sensor with 737 × 575 active pixels, where the pixel size is 6.6 μm × 6.3 μm. Figure 13-4 shows the pixels transition of the three levels of palmprint features with magnification, including principal line magnified, wrinkle magnified and ridge magnified, obtained by our proposed system. Each square grayscale grid is a pixel. The brighter one has a higher grayscale value while the darker one has a lower grayscale value. From the figure, we can see that the wrinkle and principal line are very clear with higher contrast from the background so that they can be captured easily. For the ridge-valley pattern, it occupies about 5 pixels in the image plane. The slanted pattern reduces the contrast also. For the strong ridge-valley pattern, they can be obtained close to the original signal. However, for those blurred ridge-valley patterns, the contrast is too weak to distinguish.

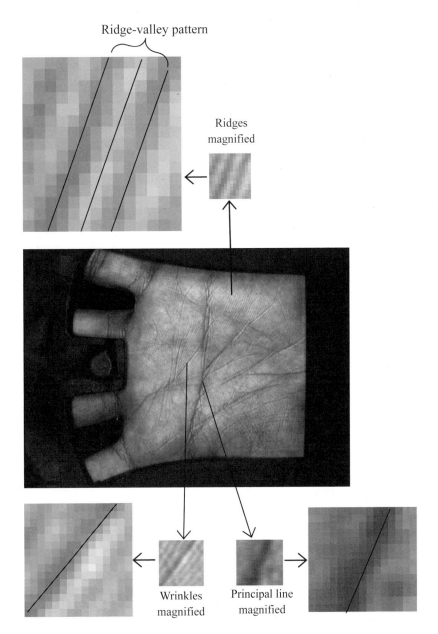

Figure 13-4. Pixels transition of the three levels of palmprint features.

In summary, the performance would be degraded by the following factors:
1) The pattern is out-of-phase with the sensor grid. Actually, this case occurs more than in-phase.
2) It is impossible to have 100% contrasts in the ridge-valley pattern. Also, some shallow ridge-valley pattern is very blur and off low contrast.

3) Usually, the ridge-valley pattern crosses the sensor grid in a slanted direction so that the signal will be degraded more, i.e. similar with the case of out-of-phase.

4) Noises and distortions are possible to be added to the image, from different parts of the whole acquisition system, including lens aberration, diffraction, fixed pattern noise from the sensor, sampling and quantization errors, etc.

As mentioned before, it is not feasible to use too high resolution sensor for a palmprint since it involves large image size and higher computation burden. By using this system configuration, we can obtain the principal lines, wrinkles, and ridges as texture feature effectively.

Resolution of the Optical System

For the performance of an optical system, people usually refer to the term resolution. There are four different types of resolutions [114-115]. They are: 1) temporal, 2) grayscale, 3) spectral, and 4) spatial.

Temporal resolution is the time of sampling frequency on an imaging system, i.e. a system operating at 25 Hz sampling rate has a temporal frequency of 40 *ms*. Grayscale resolution refers to the ability to discriminate the smallest brightness changes in the scene from the quantization process. The spectral resolution is the spectral band pass of the system, such as visible or infrared signal. The spatial resolution describes the ability of an imaging system to discriminate between two closely spaced objects, or fine details in a scene in its field of view. It involves the quantization of the continuous image at the sensor plane, to determine the digital brightness values from each grid at the sensor to form the digital image of the object scene. Let the horizontal center-to-center spacing of the sensor element array is p, the spatial Nyquist rate in each dimension is $f_{Nyquist}$, measured in line pairs per mm (lp/mm). As $f_{Nyquist} = 1/2p$, and the proposed sensor has $p = 6.6 \ \mu m$ so that $f_{Nyquist} = 75$ lp/mm. Table 13-2 shows the four types of resolution required by our system, and the respective selection reasons. There are other optical parameters such as the field of view, the depth of field, focal length, type of lens, and so on need to be precisely determined for the best performance of the optical system for the palmprint acquisition. But we are not going to discuss them in details here.

Table 13-2. Four types of resolutions.

	Resolution	Description/Reasons
Temporal resolution	40 ms	It is enough for the real time applications.
Grayscale resolution	8 bits	There are 256 grayscale levels which is enough to represent the fine details of a palmprint image. Most of the fingerprint sensors use 8 bits grayscale image. Also, > 8 bits make the image so large, increase the computation burden, and higher cost.
Spectral resolution	Visible range	We only use the visible range of the light and filter out the infrared portion which may cause noise to the image.
Spatial resolution	75 lp/mm	Precise enough for obtaining the principal lines, wrinkles and ridge texture.

User Interface Design

The acceptability of users is an essential factor for selecting a biometric technology. We pay special attention and efforts on the design of different components in the user interface, including:

i) A panel to hold a palm so that the user feels comfortable on the acquisition process.

ii) Easy guides to position the palm on the device.

iii) An appropriate size for the user interface.

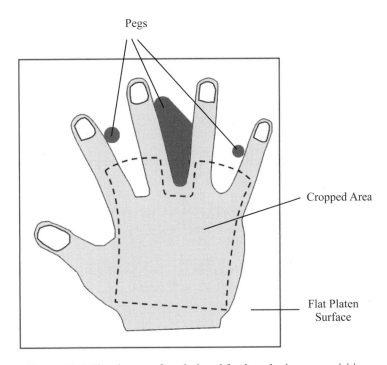

Figure 13-5. Flat platen surface designed for the palm image acquisition.

We designed a novel user interface platform for palmprint acquisition called "flat platen surface". This design makes the palm surface to be imaged by the system without contact so as to eliminate distortion caused by traditional scanners. There are some pegs on the flat platen surface to provide better guidelines for correct palm placement. This version of interface design is improved from the third version (details can be found on Chapter 2). The large triangular peg is used to guide the middle finger and the ring finger while the round-shaped pegs are used to guide the index finger and the little finger, as shown in Figure 13-5. The dotted line area is cropped from the flat platen surface to allow the CCD camera underneath to acquire the palmprint image. Under current design, we can obtain images from small or large hands appropriately. From the user's perspective, we provide an easy-to-use interface; from the system's perspective, we provide a stable positioning mechanism for palmprint alignment. The design of the flat platen surface thus satisfies both user and system perspectives.

13.3   System Performance Evaluation

This section is going to evaluate the performance of our system, including real time processing, image quality, effective palmprint pixels and a real world application example.

Real Time Processing
We wrote a program that would grab a frame of a palmprint image from our proposed device and transfer it to the main memory. The time for capturing an image is calculated by: *time to obtain one frame* + *time to write it to the main memory* + *latency time of the CPU*. The time to obtain one frame is 1 *second* / 25 *fps* = 0.04 *second*. We set up an experiment to calculate the time taken from the start of the acquisition process until the image is transferred to the main memory.

   We performed the experiments using three different speed CPUs that were connected to a standard hard disk. The total time to store a palmprint image is 0.22s, 0.11s and 0.09s on CPUs of Pentium 233 MHz, Celeron 950 MHz and Pentium 4 1.8 GHz respectively. Since the Pentium 233 MHz CPU is installed on an industrial board with an extra disk-on-chip module, we performed an extra test using DiskOnChip 2000®, and the time used was 0.11s, half of that of using a normal hard disk. This implies that the disk speed is also an important factor in image acquisition speed. We can say that our system can obtain a palmprint image in real time, even operating in a slow computing environment. On the other hand, for comparison purpose, we acquired a grayscale image from a scanner with resolution of 150 dpi using a computer running on Windows XP with Pentium 500 MHz CPU. It took ten seconds for acquiring an image of 874 × 880 in size.

Image Quality
We first analyzed the features on a palm, and how those features can be acquired by our proposed system. It is independent of algorithm, i.e. the system is by no means designed for a special algorithm for palmprint matching. Different researchers would like to use different palm features for their researches. However, we would like to show that our system is able to obtain all principal lines, wrinkles and ridge textures. Figure 13-6 (a) shows a palmprint with different resolutions: 75 dpi, 100 dpi, 125 dpi, and 150 dpi. The finest level of features (ridge texture) started to blur as the resolution decreased. Based on the collected 9,400 images, we summarized the findings on the resolution requirement for different levels of features from a palm in Figure 13-6 (b). The resolution from 125 dpi to 150 dpi is capable of obtaining all those three levels of features from a palm, whereas the resolution from 100 dpi to 125 dpi can only get wrinkles and principal lines clearly. For a resolution of less than 100 dpi, only principal lines are still clear and favorable for feature extraction. Some different palmprints with size of 420 × 420 cropped from the original images are illustrated in Figure 13-7.

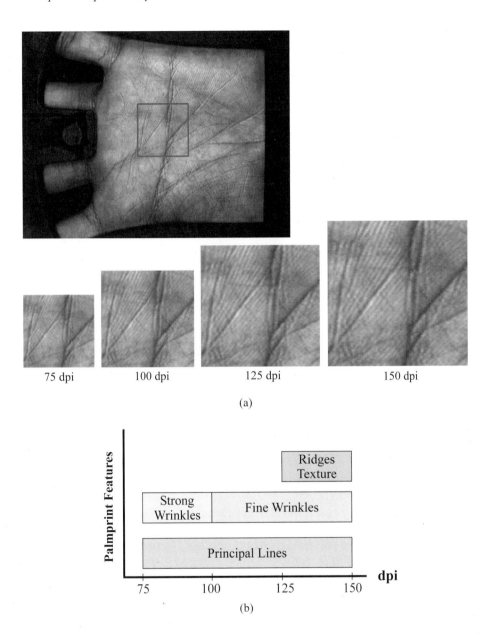

Figure 13-6. A palm with different resolutions (75 dpi, 100 dpi, 125 dpi, and 150 dpi) in (a), and the resolution requirement for different levels of features from a palm in (b).

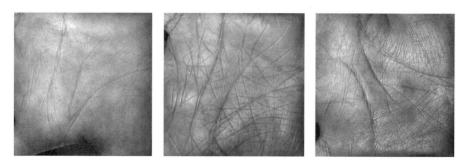

Figure 13-7. Different palmprint images.

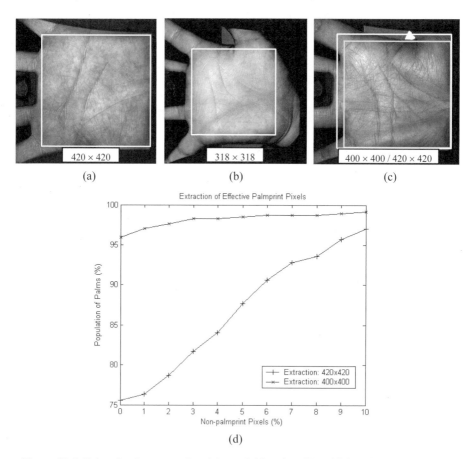

Figure 13-8. Palmprint data extraction: (a) an adult's palm, (b) a child's palm, (c) difference between extraction area of 400×400 and 420×420, and (d) experimental results on the extraction of effective palmprint pixels from different palms.

Effective Palmprint Pixels

In an ideal situation, our system is able to obtain the palmprint data (functional for palmprint feature extraction) equivalent to 420×420 pixels from the flat platen surface. However, some non-palmprint pixels are acquired from the system due to a smaller hand, leading to a smaller extracted palmprint. We carried out an experiment to examine effective palmprint pixels which can be extracted from our database. The database has a total of 9,400 images obtained from 235 individuals using our palmprint acquisition system. The subjects are mainly students and staff volunteers from The Hong Kong Polytechnic University. 157 persons are male and 78 persons are female. The age distribution of the subjects is: 10% younger than 20 years old, 78% of 20-29 year-olds and 12 % of older than 30. The results are summarized in Figures 13-8. If a palm completely covers the cropped area (dotted line) on the flat platen surface, our system is able to get a full-size of 420×420 pixels of palmprint data, as shown in Figure 13-8 (a). However, we could only get as much as 318×318 pixels (a number smaller than the expected values) from a child's hand, as shown in Figure 13-8 (b). From Figure 13-8 (d), it can be seen that by using a 420×420 pixels area, only 75% of palms can obtain palmprint data without non-palmprint pixels. On the other hand, by using a 400×400 pixels area, more than 95% of palms can obtain palmprint data without non-palmprint pixels. The difference is shown in Figure 13-8 (c) where the outer white box is the extraction area equivalent to 420×420 pixels, containing some non-palmprint pixels that occurred on the top and the left of the image. In the inner box, an area of 400×400 pixels, we obtained no non-palmprint pixel. This experimental result suggests that our system is effective on obtaining palmprint data at 400×400 pixels with more than 95% of palms without non-palmprint pixels.

Real World Example

To test the feasibility of our palmprint acquisition system, we connected it to a recognition engine developed by us [65]. We put them together to form a palmprint identification system which is placed at our laboratory entrance for access control since early March 2003. Details can be found at Chapter 14. The palmprint identification system is able to store 400 users, with identification time less than 1.5 seconds. Details of the recognition engine can be found at [65]. All the personnel inside the laboratory have enrolled to the system. We have been using our palm as the 'key' to enter the laboratory.

13.4   Summary

In this chapter, we introduced novel ideas for palmprint acquisition including a specially designed user interface, the real-time palmprint image acquisition, and high quality image with reasonable price and relatively small in size. We performed a requirement analysis on the proposed system in terms of the minimum feature size, spatial resolution and user interface. During the implementation stage, we optimized the optical parameters in order to achieve the best performance in terms of palmprint image quality and size of the system. The fastest time to obtain a palmprint image at spatial resolution of 150 dpi is 0.09s. Our proposed system is able to obtain features from a palm including principal lines, wrinkles and ridge textures. For the user interface, a special designed platform called the flat platen surface was built to guide the placement of palms during the acquisition process. The novel contactless design on the user interface prevents the distortion of the palm skin caused by the traditional glass plate of a palm scanner. The proposed system provides an important step for implementing a personal identification system using palmprint. Last but not least, we have implemented the system at our laboratory entrance for access control since early March 2003 to demonstrate the viability of the system. We have concluded that our system is effective for palmprint acquisition in real time with high image quality and low cost.

14 A PROTOTYPE DESIGN FOR CIVIL APPLICATIONS

In this chapter, we present a biometric authentication system for identifying a person by his/her palmprint. We first briefly introduce the need for the development of a palmprint authentication system for civil applications. Then, in Section 14.2 we give an overview of the system and describe its components. Section 14.3 reports the performance evaluation of the proposed system, including experimental results of verification and identification. Section 14.4 discusses different issues related to the proposed system, such as the robustness measures, computation power and time requirements, and extensibility of the system. Section 14.5 discusses potential applications of our proposed system. Section 14.6 summarizes this chapter.

14.1   Introduction

There has been an ever-growing need to automatically authenticate individuals at various occasions in our modern and automated society, such as physical access control on building entrance and logical access control on computer logon. Traditional knowledge based or token based personal identification or verification is so tedious, time-consuming, inefficient or expensive, which is incapable to meet such a fast-pacing society. Knowledge-based approaches use "something that you know" to make a personal identification, such as password and personal identification number. Token-based approaches use "something that you have" to make a personal identification, such as passport or credit card. Since those approaches are not based on any inherent attributes of an individual to make the identification, it is unable to differentiate between an authorized person and an impostor who fraudulently acquires the "token" or "knowledge" of the authorized person. This is why biometrics identification or verification system started to be more focused in the recent years.

In fact, using biometrics for person authentication is not new, which has been implemented over thousands years. Undoubtedly, face recognition is a direct and intuitive approach for human beings as the most common biometric. In addition, fingerprint and signature are two important biometrics technologies, which have been used to approbate the contents of a document or to authenticate a financial transaction. The proposed palmprint system is a hand-based biometric technology, on the other hand, exploiting the features in the inner surface of our palm for personal identification. Therefore, we expect that the palmprint system will receive a high user acceptance, like fingerprint, hand geometry and hand vein [6, 8, 16, 118]. Because of the rich features including texture, principal lines and wrinkles on palmprints, we believe that they contain enough stable and distinctive information for separating an individual from a large population. All in all, we are sure that palmprint has the strengths of stability, reliability, convenience and comfortable to use.

1) *Stability* – The chance of getting damage of a palm is lower than that of fingerprint. The line features of a palm are not easy to change in a considerable long period of time.

2) *Reliability* – The surface areas of a palm is very large so that many unique features can be extracted to represent a person leading to a more reliable identification and verification result. In addition, small dirt or grease on the palm of a user is not sensitive for palmprint identification as we have got a large palm area with lower image resolution than fingerprint images. Those dirt or grease only shares a little percentage of the whole palm.

3) *Convenience* – By our dedicated design on the user interface, it is convenient to use by putting ones hand on the flat platen surface of the palmprint device, it is really a simple act.

4) *Comfortable to use* – Most of people associate the fingerprint authentication with criminality and feel unease of using it. With palmprint, there is no such connection and people may feel comfortable to use it.

There is no commercial product available on the market for the palmprint authentication for civil applications, we think that there is a need to develop this new type of biometric device and system, as a new member of the biometrics family. On the basis of our previous research work [65], we developed a novel palmprint authentication system to fulfill such requirements. According to our best knowledge, the prototype reported in this chapter is the first real-time palmprint identification system for commercial and civilian applications in the world.

14.2   System Framework

The proposed palmprint authentication system has four major components: they are *User Interface Module, Acquisition Module, Recognition Module* and *External Module*. Figure 14-1 shows detailed components of each module of the palmprint authentication system. Figure 14-2 gives the palmprint authentication system installed at the Biometric Research Center, Department of Computing, The Hong Kong Polytechnic University. The functions of each component are listed below:

A) *User Interface Module* provides an interface between the system and users for the smooth authentication operation. We designed a flat platen surface for the palm acquisition (described in details in Chapters 2 and 13). It is crucial to develop a good user interface such that users are happy to use the device. Also, there are some servicing functions provided for the administrator, such as enroll user, delete user, etc. A LCD displays appropriate message to instruct the users in the enrollment process and during/after the identification process.

B) *Acquisition Module* is the channel for the palm to be acquired for the further processing. It calls the frame grabber to transfer one frame of image to the processor, and then examines whether a hand has put on the device (described in details also in Chapters 2 and 13).

C) *Recognition Module* is the key part of our system, which determines whether a user is authenticated. It consists of image preprocessing to detect the key points of the hand for central part extraction, feature extraction to obtain some effective features, template creation to store the registered features (during enrolment), database updating to add the new user information (during enrolment), and matching to compare the input features and templates from the database (on the identification process).

D) *External Module* receives the signal from the recognition module, to allow some operations to be performed, or deny the operations requested. This module actually is an interfacing component, which may be connected to another hardware components or software components. Our system uses a relay to control an electronic door lock. On the other hand, if the proposed system is employed on an employee attendance system, it will keep track of the record on who has identified at what time, and finally send this information to the respective accounting system.

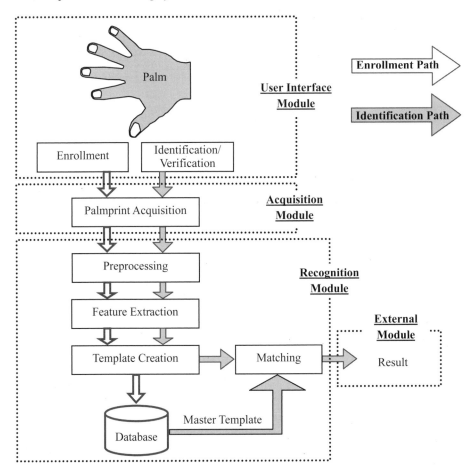

Figure 14-1. Four modules of the palmprint authentication system.

Figure 14-2. A palmprint identification system at a laboratory entrance for access control.

The core of the recognition module is modified from [65], composed of images testing, preprocessing, feature extraction and matching. This prototype has considered the automatic image collection and image selection process. In addition, the feature extraction of the proposed system has a great improvement.

Image Selection
An automatic image collection is an important part of the system, which is used to detect potential images for palmprint recognition. In the image selection, image difference is used to select the potential images. Originally, if nothing moves on the palmprint acquisition device, the image difference of two images should be small. When a user put his/her hand on the palmprint acquisition device, it causes relatively large image difference at the beginning because of the movement. Once the user rest his/her hand on the flat platen surface of the acquisition device, the corresponding image difference becomes small. Based on these properties, the following two inequalities are applied to select potential images,

$$\sum_{y=o}^{Y}\sum_{x=0}^{X}\left|I_i(x_o+hx,y_o+vy)-I_j(x_o+hx,y_o+vy)\right| > S_1,$$

(14-1)

and

$$\sum_{y=o}^{Y}\sum_{x=0}^{X}\left|I_p(x_o+hx,y_o+vy)-I_q(x_o+hx,y_o+vy)\right| < S_2,$$ (14-2)

where I_i, I_j, I_p and I_q are the images from a sequence of images, I_k, $k=0,\ldots,N$; h and v control the horizontal and vertical sampling, respectively; (x_0, y_0) is the starting point of the sampling region; $h{\times}X$ and $v{\times}Y$ are the height and width of the sampling region, respectively. Since only the central part of the images would be palmprint pixels, $h \times X$ and $v \times Y$ do not equal to the width and height of the input image, respectively. If the image sequence fulfills Equations 14-1 and 14-2, the last image I_q is sent to the palmprint preprocessing component.

Preprocessing
When capturing a palmprint, the position, direction and stretching degree may vary from time to time. As a result, even the palmprints from the same palm could have a little rotation and translation. Also, the sizes of palms are different from one another. So, the preprocessing algorithm is used to align different palmprints and extract the corresponding central part for feature extraction [119]. In our palmprint identification system, both the rotation and translation are constrained to some extent by the flat platen surface of the acquisition device, which can guide the palms by the pegs. Then the preprocessing algorithm described in Chapter 6 Section 6.3 can be used to locate the coordination system of the palmprints effectively.

Feature Extraction
The feature extraction technique implemented on the proposed palmprint identification system is modified from [65], where single circular zero DC Gabor filter is applied to the preprocessed palmprint images and the phase information is coded as feature vector called PalmCode. Details information on the feature extraction can be found in Chapter 7 Section 7.1.

Matching
The feature matching determines the degree of similarity between two templates – the authentication template and the master template. Details information on the feature matching also can be explained in Chapter 7 Section 7.2.

14.3 Experimental Results

We use all the images collected from the **PolyU-ONLINE-Palmprint-II**, which has 200 individuals, obtained by our palmprint capture device described in [65]. There are totally 8,025 palmprints from 400 different palms in our database. The maximum and the minimum time intervals between the first and second palmprint acquisition were 162 days and 1 day, respectively. The details of the database information are given in Chapter 2 Section 2.3.

Experimental Results of Verification
Verification or authentication refers to the problem of confirming or denying a claim of individuals (Am I who I claim I am?). It is also considered as one to one matching, one person to one person matching. Verification only limits the number of persons involving in the test but does not limit number of images or trials involving in the test. Therefore, two common practices have been implemented. The first one only involves two images, one for registration, one for testing, which is only suitable for template matching. Therefore, complex classifiers such as neural networks and

support vector machine requiring more samples for training the system cannot be tested under the first approach. However, the first approach is commonly used to test the performance of fingerprint systems [59, 120-121]. The other approach involves many samples in registration for training the system [122-125], which is useful for face and signature verifications. In fact, many commercial biometric verification systems, including fingerprints store several templates of a user's fingerprint for obtaining better system performance. To compare the verification results of fingerprint and hand geometry verification systems reported by academic and industry [6, 59, 120, 126], two approaches are implemented for testing our palmprint identification system.

To obtain the verification result of the first approach, each palmprint image is matched with all other palmprint images in the database. A matching is counted as a correct matching if the two palmprint images are from the same palm; otherwise, the matching is counted as incorrect matching. The total number of matching is 32,119,735. Number of genuine matching is 76,565 and the rest of them are incorrect matchings. Figure 14-3 (a) shows the probability of genuine and imposter distributions estimated by the correct and incorrect matchings, respectively and Figure 14-3 (b) illustrates the corresponding Receiver Operating Characteristic (ROC) curves, as a plot of the genuine and acceptance rate against the false acceptance rate for all possible operating points. Some thresholds and corresponding false acceptance rates (FAR) and false rejection rates (FRR) are listed in Table 14-1. According to Table 14-1, using one palmprint image for registration, the proposed palmprint system can be operated at low false acceptance rate, 0.096% and reasonable false rejection rate, 1.05%.

To obtain the verification accuracy of the second approach, the testing database is divided into two databases, 1) registration database and 2) testing database. Three palmprint images of each palm collected in the first occasion are selected for the registration database. Totally, the registration database contains 1,200 palmprint images and the rest of them are for the testing database. In this verification test, each palmprint image is matched with all the palmprint images in the testing database. Therefore, each testing image can produce three hamming distances for one registered palm. The minimum of them is regarded as the final hamming distance. For achieving statistically reliable results, this test is repeated three times by selecting other palmprint images for the registration database. Total number of hamming distances from correct matchings and incorrect matchings are 20,475 and 8,169,525, respectively. Figure 14-3 (c) shows the probability of genuine and imposter distributions estimated by the correct and incorrect matchings, respectively and Figure 14-3 (d) indicates the corresponding ROC curves. Some thresholds and the corresponding false acceptance and false rejection rates are also listed in Table 14-1. According to both Table 14-1 and Figures 14-3, we can conclude that using three templates can provide better verification accuracy. In fact, more palmprint images of the same palm can provide more information to the system so that it can recognize the noise or deformed features. It is also the reason for commercial verification systems requiring more than one biometric signal, such as fingerprint images for registration.

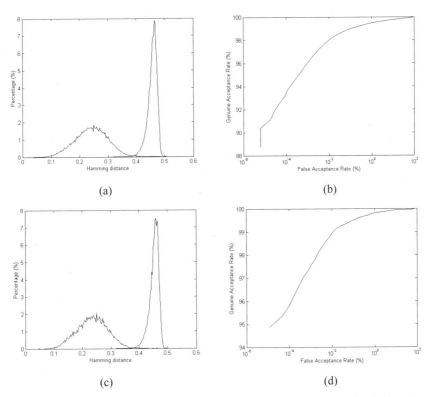

Figure 14-3. Verification test results. (a) and (b): genuine and imposter distributions for verification tests with one and three registered images per palm, respectively. (c) and (d): the corresponding receiver operating characteristic (ROC) curves of (a) and (b), respectively.

Table 14-1. False acceptance rates (FARs) and false rejection rates (FRRs) with different threshold values for the palmprints verification results.

Threshold	No. of registered image = 1		No. of registered image = 3	
	FAR (%)	FRR (%)	FAR (%)	FRR (%)
0.32	0.000027	8.15	0.000012	5.12
0.34	0.00094	4.02	0.0016	2.18
0.36	0.011	1.94	0.017	0.86
0.38	0.096	1.05	0.15	0.43
0.40	0.68	0.59	1.03	0.19

Experimental Results of Identification

Identification test is a one-against-many, N comparison process. It is also the target of our palmprint identification system for identifying a palm in a relatively large database. In this experiment, N is set to 400, which is the total number of different palms in our database. Same as the previous verification experiment, the palmprint database are divided into two databases, 1) registration and 2) testing databases. The registration database contains 1,200 palmprint images, three for each palm and the testing database has 6,825 palmprint images. Each palmprint image in the testing database is matched with all the palmprint images in the registration database. Therefore, each testing image generates 3 correct and 1,197 incorrect matchings. The minimum hamming distances of correct matchings and incorrect matchings are regarded as identification hamming distances of genuine and imposter, respectively. This experiment is also called one trial test since the user only provides one palmprint image in the test to make one decision. In fact, a practical biometric system collects several biometric signals to make one decisions. Therefore, in this experiment, we implement one, two and three trial tests. In the two trials testing, a pair of palmprint images in the testing database belonging to the same palm is matched with all the palmprint images in the registration database. Each pair of palmprint images in two trial test generates 6 correct and 2,394 incorrect matchings. Similarly, in the three trials test, the identification hamming distances of genuine and imposter are obtained from 9 correct and 3,591 incorrect matchings, respectively. Each test is repeated three times by selecting other palmprints for registration database. In each test, the number of identification hamming distances of genuine and imposter matchings both are 20,475. Figure 14-4 shows ROC curves of the three tests and Table 14-2 lists the thresholds and the corresponding FAR and FRR. According to Figure 14-4 and Table 14-2, more input palmprints can provide more accurate results.

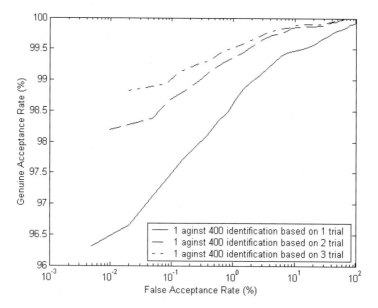

Figure 14-4. The ROC curves are for a 1-against-400 identification testing with different number of trials.

Table 14-2. False acceptance rates (FARs) and false rejection rates (FRRs) with different threshold values for the 1-to-400 palmprints identification results.

Threshold	Trial = 1		Trial = 2		Trial = f3	
	FAR (%)	FRR (%)	FAR (%)	FRR (%)	FAR (%)	FRR (%)
0.320	0.0049	3.69	0.0098	1.80	0.020	1.17
0.325	0.0439	2.93	0.088	1.34	0.131	1.06
0.330	0.15	2.29	0.28	1.02	0.42	0.68
0.335	0.37	1.90	0.68	0.72	0.96	0.48
0.340	0.84	1.51	1.43	0.57	1.93	0.37
0.345	1.45	1.16	2.32	0.42	3.02	0.26

14.4   Performance Evaluation

As a practical biometric system for civil applications, other than accuracy, robustness of the system, computation power and time requirements, extensibility of the system, and cost of the system, are also important issues. In this section, we will discuss these issues one by one.

Robustness
There are some situations which affect the performance of a biometric system, grouping into two categories: 1) noises from the user's hand or from the environment, and 2) special characteristics of the user's palmprint. Some experiments are performed to illustrate the robustness of our system.

Category 1: noises from the user's hand or from the environment
Three common noises sometimes appearing on the hand should be considered. The first one is an individual's jewelry such as rings, which may influence the accuracy of some preprocessing algorithms. The second one is the noise such as text wrote on the palm, which directly affects the performance of the system. The third one is the varying lighting condition from the environments. Although our palmprint acquisition device provide a semi-closed lighting environment, it is still affected by some outside lighting.

   Three experiments are designed to test the robustness of the proposed system. They are: (1a) palmprint images with rings (see Figure 14-5), (1b) palmprints with text (see Figure 14-6), and (1c) palmprint images collected under very different lighting environments (just as shown in Figure 14-7).

1a. Palmprint images with rings
Figures 14-5 show three palmprint images, with and without wearing ring(s) on the fingers and their corresponding preprocessed images. We can see that our preprocessing algorithm described in Chapter 6 Section 6.3 is not affected by the rings since it does not exploit the information on the boundary of the finger. However, some preprocessing algorithms such as [80] using such information may not be stable under the influence of the rings. From Table 14-3, the hamming distances of genuine

matchings are relatively small, i.e. less than 0.26. That is, the proposed palmprint identification system is robust to the fingers with rings.

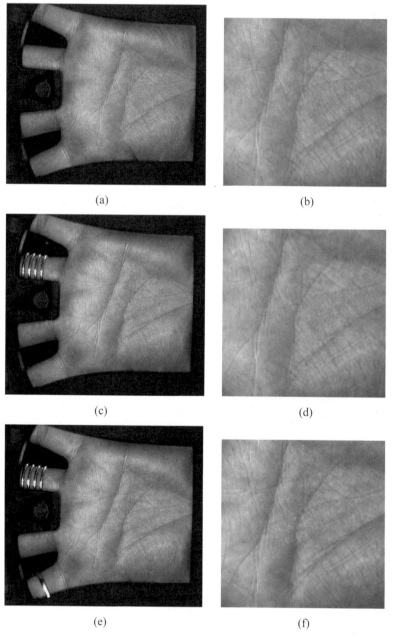

(a)

(b)

(c)

(d)

(e)

(f)

Figure 14-5. Palmprints with rings. (a), (c), (e): original images; (b), (d), (f): central part palmprint sub-images.

Table 14-3. The hamming distances from genuine matchings of Figures 14-5.

Figures	14-5 (c)	14-5 (e)
14-5 (a)	0.24	0.25
14-5 (c)		0.18

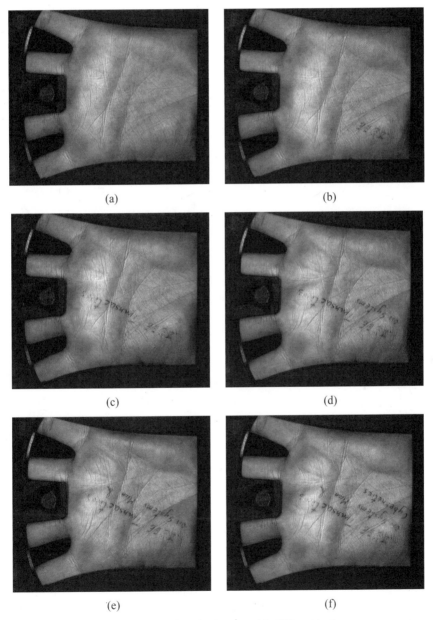

(a)　　　　　　　　　　　　　(b)

(c)　　　　　　　　　　　　　(d)

(e)　　　　　　　　　　　　　(f)

Figure 14-6. Palmprint images with different text.

Table 14-4. The hamming distances from genuine matchings of Figures 14-6.

Figures	14-6 (b)	14-6 (c)	14-6 (d)	14-6 (e)	14-6 (f)
14-6 (a)	0.19	0.21	0.27	0.29	0.28
14-6 (b)		0.18	0.27	0.26	0.27
14-6 (c)			0.27	0.28	0.28
14-6 (d)				0.23	0.19
14-6 (e)					0.19

1b. Palmprints with text

Figure 14-6 (a) provides a clear palmprint image while Figures 14-6 (b)-(f) show five palmprint images, with different text. Their hamming distances are given in Table 14-4; all of them are smaller than 0.3. Comparing to the hamming distances of imposter in Tables 14-1 and 14-2, it demonstrates that all the hamming distances in Table 14-4 are relatively small. That is, the proposed palmprint identification system is robust to the noise caused by the written text on the palmprint.

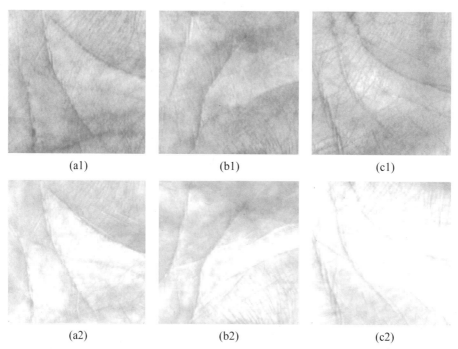

(a1)	(b1)	(c1)
(a2)	(b2)	(c2)

Figure 14-7. Preprocessed palmprint images under different lighting environments. (a1), (b1), (c1): normal condition; (a2), (b2), (c2): varied lighting condition.

1c. Palmprint images collected under very different lighting environments

Figures 14-7 show some palmprint images obtained in two different occasions, with different lighting conditions. Some of the fine details are blurred by this reason. From the hamming distances given in Table 14-5, the highest genuine matching distance is

0.36 while the imposter matching distances are around 0.44-0.48. That means we still can separate them from each others properly.

Table 14-5. The hamming distances of Figures 14-7; the highlighted hamming distances are obtained from genuine matchings while the remainings are from imposter matching.

Figures	14-7 (a2)	14-7 (b1)	14-7 (b2)	14-7 (c1)	14-7 (c2)
14-7 (a1)	0.27	0.48	0.46	0.48	0.46
14-7 (a2)		0.47	0.44	0.47	0.44
14-7 (b1)			0.32	0.47	0.48
14-7 (b2)				0.48	0.47
14-7 (c1)					0.36

Category 2: special characteristics of the user's palmprint

Apart from the noise problems depicted in Category 1, the special characteristics of a user's palmprint can affect the accuracy of a system. For example, sometimes a user with very similar line features may cause a false match while a palmprint without clear line feature could make the system failed to enroll a person.

Three experiments are designed to test the robustness caused from this category, including: (2a) palmprints with similar line features (see Figure 14-8), (2b) palmprint without clear line features (see Figure 14-9), and (2c) identical twins' palmprints (see Figure 14-10).

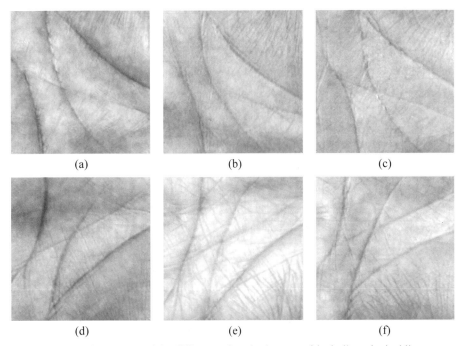

(a)   (b)   (c)

(d)   (e)   (f)

Figure 14-8. Two sets of the different palmprint images with similar principal lines.

Table 14-6. The hamming distances from imposter matchings of Figures 14-8.

Figures	14-8 (a2)	14-8 (b1)	14-8 (b2)	14-8 (c1)	14-8 (c2)
14-8 (a1)	0.44	0.44	0.45	0.45	0.47
14-8 (a2)		0.47	0.44	0.47	0.46
14-8 (b1)			0.48	0.47	0.46
14-8 (b2)				0.40	0.41
14-8 (c1)					0.44

2a. Palmprints with similar line features

Figure 14-8 provides some palmprint images with similar line features from different persons. Their hamming distances are given in Table 14-6; all of them are larger than 0.39 which is enough to separate different persons.

2b. Palmprint without clear line features

Figure 14-9 indicates some palmprint images without clear line features. From the hamming distances given in Table 14-7, the highest genuine matching distance is only 0.29 while the imposter matching distances are around 0.45-0.47. That means we can easily to separate them from each others properly.

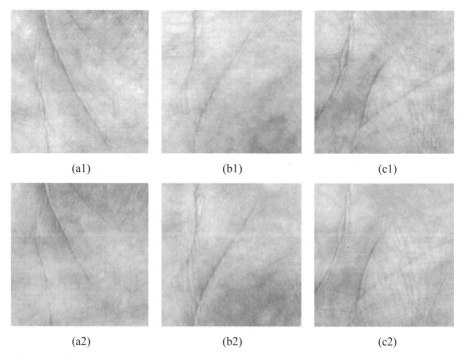

(a1)   (b1)   (c1)

(a2)   (b2)   (c2)

Figure 14-9. Three pairs of palmprint images without clear wrinkles. (a1, (b1), (c1): first time collected images; and (a2), (b2), (c2): second time collected images.

Table 14-7. The hamming distances of Figures 14-9; the highlighted hamming distances are obtained from genuine matchings while the remainings are from imposter matching

Figures	14-9 (a2)	14-9 (b1)	14-9 (b2)	14-9 (c1)	14-9 (c2)
14-9 (a1)	0.28	0.46	0.47	0.47	0.46
14-9 (a2)		0.47	0.47	0.47	0.47
14-9 (b1)			0.29	0.46	0.46
14-9 (b2)				0.45	0.46
14-9 (c1)					0.24

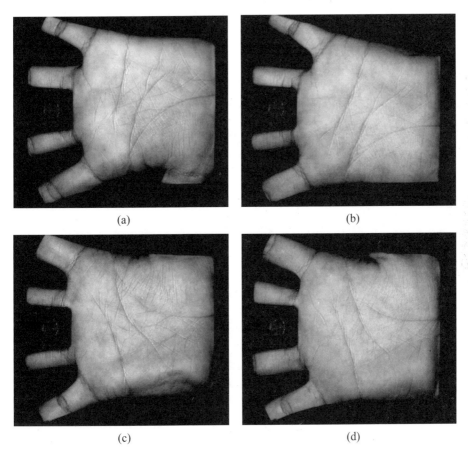

(a)   (b)

(c)   (d)

Figure 14-10. Identical twins' palmprints. (a), (b) and (c), (d) are their left and right hands, respectively.

2c. Identical twins' palmprints

A test of identical twins is regarded as an important test for biometric authentication system, but not all biometrics, including face and DNA, can pass this test. However, the palmprints of identical twins have enough distinctive information to distinguish them. We collected 590 palmprint images from 30 pairs of identical twins' palms. Some samples are shown in Figures 14-10. Each of them provides around 10 images of their left palm and 10 images of their right palm. Their age range is between 6 and 45 years old. Based on this database, we match a palmprint in the twin database with his/her identical twin sibling to produce imposter matching scores, and match the samples of their own to get the genuine scores. From the hamming distances given in Table 14-8, the lowest imposter matching distance is 0.41 which means that our algorithm is robust to separate them from each others properly. The genuine and imposter distributions are given in Figure 14-11. From the figure, we can find that identical twins' palmprint can easily be separated, just like twins' fingerprints [127].

Table 14-8. The hamming distances from imposter matchings of Figures 14-10.

Figures	14-10 (b)	14-10 (c)	14-10 (d)
14-10 (a)	0.42	0.46	0.45
14-10 (b)		0.46	0.46
14-10 (c)			0.41

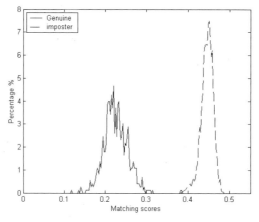

Figure 14-11. The genuine and imposter distributions for measuring the similarity of identical twins' palmprints.

We have performed six experiments, targeted on the testing of issues like: palmprints images with rings, palmprints with text, palmprint images collected under very different lighting environments, palmprints with similar line features, palmprint without clear line features, and identical twins' palmprints. Experiment results show that our proposed system has high robustness, from palmprint preprocessing, feature extraction to feature matching, which can face different challenges in the real world situations. We can say that our proposed system is comparable with other hand-based biometrics systems, such as hand geometry and fingerprint verification system [6, 8, 59] and can be used for real-world applications. As mentioned before, the proposed palmprint identification system has been installed at the entrance of our laboratory since March 2003 for access control, see Figure 14-2.

Computation Time

Another key issue for a civilian personal authentication system is whether the system can run in real time. In other words, the system should be running as fast as possible. The proposed method is implemented using C language and Assemble language on a PC using Intel Pentium IV processor (1.4GHz) with 128MB memory. The execution time for image collection, image preprocessing, feature extraction and matching is listed in Table 14-9. The total execution time for a 1-against-400 identification, each palm with 3 templates, is less than 1s. Users will not feel any delay when using our system.

Actually, we did not optimize the code so it is possible to reduce the computation time further. For example, if the code is rewritten using all the assembly language running on the DOS environment, further reduction in the computation time is expected.

Table 14-9. Execution time of the palmprint authentication system.

Operations	Execution Time
Image collection	340ms
Preprocessing	250ms
Feature extraction	180ms
Matching	1.3μs

Extensibility of Our System

The primary design objective of our system is to be a personal identification system for civilian applications; however, by adding a token such as smart card or ID number, it is not difficult for our system to operate in verification mode to become a personal verification system. That is, the final system is an identification or verification system highly dependent on the implementation of the users, company and application domains.

In fact, our system is a generic platform which provides the basic functionality on identification/verification of a person, but the output interface is flexible enough to be replaced by the hardware and/or software modules to achieve a specific requirement. In case of using our system as an employee attendance management system, there maybe no 'physical' gateway for the employees to go through, rather, the employee should put their palm down on the flat platen surface for attendance tracking when they are on-duty and off-duty. A corresponding software module should keep track on those attendance record for the purpose of calculating the salary of each employee.

Currently, our system is a standalone device which is connected to a physical control application. The external module sends an electrical signal to an electrically controlled door to unlock the door when the identification is success. By the way, we can extend our system to operate with multiple authentication points by placing the capture devices on different authentication points and connecting by network cables to provide a complete solution to those possible requirements.

Cost of Our System

With the advances in silicon technology, the chips become lower price but higher quality. For a low to medium profile black and white CCD sensor board, the price is less than US$100. Our proposed device will use a high quality lens which costs about US$35. The cost of fluorescent light and video frame grabber altogether is less than US$50. Some common hardware components for the computation work include a motherboard, a CPU of Intel Pentium 1.4 GHz, main memory of 128 MB RAM and storage of 32MB Disk on Module for the storage of users database. All those hardware components can be bought from the market easily with less than US$300. Summing all the costs of the above components, the total cost is less than US$500. The cost stated here does not include the assembling, wiring, cover case, etc.

According to the marketing information, 3D Hand Geometry System usually cost around US$2,500. Both of us are hand-based biometric technologies, our system has definite advantage compared with 3D Hand Geometry System in terms of the price. In addition, our system can perform 1-to-400 user identification which does not need to use with smart card, PIN number or ID number.

14.5   Potential Applications

Biometrics can be used in systems ranging from customer oriented to employee oriented applications such as retail banking, ATMs, airports, time attendance management, and so on, which improve the work flow effectively and eliminate frauds. Our system can be treated as a supplement of existing service or even a replacement of current method such as smart card or password based authentications. Our proposed system is an identification system, but it can also be used as a verification (authentication) system by adding a token such as smart card in some application domains such as ATMs and credit card purchases.

A.   Audit Trail and Anti-fraud Applications

Retail Banking

The internal operation of retail banking is an employee oriented applications which is suitable to implement the biometrics identification mechanism. In retail banking system, different level of authorizations are needed in daily operations such as activation of some inactive accounts, cash-in cheques which over a specified limits, etc. Some banks have implemented an authorization mechanism for different hierarchies of staff in approving different operations by swiping their badge or card for audit trail. But the supervisor badge or card can be stolen, lent to other persons, or lost, which will be difficult to trace in case of fraud. Also, it is a frequent action in a busy working day for those senior staff on the authorization process, which is inconvenience and also affects their daily work. Palmprint identification system eliminates this kind of problems by placing a device on each of those supervisors' desk for the approval process. When there is a need by the front line staff or junior staff, the requests will be passed to the supervisors' computer by a pop-up message, then he/she can approve the request by his/her palmprint, records can be tracked automatically.

ATM

Although the existing authentication environments for ATM operations are imperfect, it is the most effective and widely accepted mechanism. ATM is a customer oriented application where biometrics can be implemented to replace the traditional mechanism such as facial-scan, finger-scan and iris-scan. They use the biometric features to replace the traditional password based method to prevent fraud. By using our proposed system with a smart card, no more frauds can be occurred by the lost or theft of password along with an ATM card.

Credit Card Purchases

Credit card frauds are major problems for retailers and Credit Card Companies which cause huge losses. Some criminal groups stole the credit card information to create a new card to purchase. The implementation of a biometric verification system, such as our proposed palmprint system which stores the templates on the chips of credit card for verification, can significantly reduce fraud. It is a customer oriented application which can be used to protect financial institutes, merchant and customers. It is really convenience to consumers by putting their palm over the palmprint device during the payment process, and no signature is required anymore.

B.   Transportation Applications

Airports

There is a project called INSPASS which allows those frequent travelers use the 3D hand geometry at several international airports such as Los Angeles, Washington, and New York. Qualified passengers need to enroll the service to receive a magnetic stripe card with their hand features encoded. By implementing our proposed palmprint system, palmprint features will be stored on the ticket on enrollment process. Then the check out process is simply done by swiping their card, put their palm on the device, and proceed to the customs gate to avoid the long queuing at those airports.

C.   Physical Access Control

Building Access Control

The fundamentals of implementing a building access control is to control individuals by entering or leaving an area such as entrance of a buildings, rooms, laboratories or restricted areas by verifying their identity using our proposed palmprint system. These kinds of applications can be a customer oriented system or employee oriented system, depending on the implementation. Our system can be used to complement or replace authentication mechanisms such as keys, tokens, and badges. It can keep track of the actual user who has entered to particular areas because biometrics cannot be shared or borrowed. It creates an audit trail which is difficult to repudiate.

Time and Attendance Management

Time and attendance management is an area that needs the biometrics mechanism to detect and prevent frauds. A well known problem called buddy punching is caused by employees punching time cards for their friends who might be late or absent altogether from work. It is a headache problem for many large companies which lose hundreds of millions of dollars every year. By implementing our proposed palmprint system to replace time cards could effectively eliminates this type of frauds.

D. Other Types of Applications

Citizen ID Program
It is a trend for the governments to use biometrics technology on issuance of citizen identity cards. In Hong Kong, there is a project called Smart Identity Card System (SMARTICS) uses fingerprint as the identifiers for the authentication process. The entire population can be getting benefit as it will provide an efficient government services by this smart card with better security and faster processing time on different operations such as driver license or border crossing. An optional digital certificate can be buy and stored on the smart card to provide secure e-commerce applications. Immigration-related activities can also be getting advantages by using SMARTICS. We think that our proposed system could be adopted to provide similar services and performed as good as the fingerprint system.

Voting and Voter Registration
Voter registration and voting process is a problem in some third-world countries where some people voting twice or some cheating occurred on counting the vote tickets. By implementing our proposed system, those cheating could be eliminated. Also, the voting result can be almost instantly obtained since there is no need for people to count the votes. This type of application can be replaced by the Smart Identity Card System, if there is one.

14.6 Summary

In this chapter, we presented a novel biometric system based on the palmprint. The proposed system can accurately identify a person in real time, which is suitable for various civil applications such as access control. Experimental results show that the proposed system can identify 400 palms with a low false acceptance rate, 0.02%, and a high genuine acceptance rate, 98.83%. For verification, the system can operate at a false acceptance rate, 0.017% and a false rejection rate, 0.86%. Experimental results including accuracy, speed and robustness demonstrate that the palmprint authentication system is comparable with other hand-based biometrics systems, such as hand geometry and fingerprint verification system and is practical for real-world applications. The system has been installed at the Biometric Research Center, Department of Computing, The Hong Kong Polytechnic University since March 2003 for access control.

15 BOOK REVIEW AND FUTURE WORK

Biometric authentication, the use of the human body or human behavior for personal authentication, has a long history. In fact, we use it everyday. We commonly recognize people based on their face, voice and gait. Signatures are recognized as an official verification method in legal and commercial transactions. Fingerprints and DNA have been considered effective methods for forensic applications including investigating crimes, identifying bodies and determining parenthood. Recently, more and more effort has been put on developing effective automatic personal identification systems for various security demands. Palmprint has rich features including principal lines, wrinkles and texture, and we believe that it has stable and distinctive information sufficient for distinguishing an individual from a large population. Our team has worked on palmprint since 1996. We first worked on the offline method and built two offline palmprint databases for the algorithm design and testing. After that, we extended our work on the online real-time palmprint acquisition device. Finally, we built a large online palmprint database which allowed even more work on the algorithm using online palmprint images. Different technical papers have been published discussing the palmprint recognition algorithms including palmprint registration, line feature extraction, texture feature extraction, and transform based palmprint feature extraction. This book unveils automatic techniques for palmprint authentication, from the first approaches for offline palmprint images to the current state-of-the-art algorithms. It may serve as a handbook of palmprint authentication and be of use to researchers and students who wish to understand, participate, and/or develop a palmprint authentication system. It would also be useful as a reference book for a graduate course on biometrics.

In this chapter, we first recapitulate the contents of this book in Section 15.1, then, Section 15.2 discusses the future of palmprint research.

15.1 Book Recapitulation

This book has four parts divided into fifteen chapters which discuss the palmprint technology ranging from the hardware components of the palmprint acquisition, to the algorithm designed for the palmprint preprocessing, feature extraction and matching. Chapter One introduces recent developments in biometric technologies, some key concepts in biometrics, and the importance of developing a new biometric, the palmprint.

Chapter Two focuses on the palmprint data: offline and online palmprints. Before the online palmprint acquisition device is developed, all the palmprints are obtained offline. Offline palmprints are obtained by placing an inked palm on paper, and then using a scanner to digitize the signal. Online palmprint acquisition is the most direct way to digitize a palmprint signal. A palmprint acquisition device developed by us was revealed. Different databases are built from these two methods for testing various experiments. At last, we list the information on different palmprint databases, including the image size, number of images, resolution, etc.

Chapter Three reports two registration methods for working on offline palmprints. The first method defines and extracts datum points on a palmprint which are useful for palmprint alignment. From the experiments, 286 out of 300 palmprints are found to be in excellent agreement with the manual estimate. The second alignment method uses two invariant characteristics of a palmprint to handle the palmprints with rotation and translation. Based on this alignment method, up to 13% correct identification rate can be improved.

Chapter Four reports a line based feature extraction and matching strategy for offline palmprint verification. Based on our observations, the curvature of the principal lines is small enough to be represented by several straight line segments. Straight lines are easy to represent and match. The matching of palm lines is performed by measuring the Euclidean distances between the endpoints of two line segments and also the angles between these line segments. Then, a classification method is proposed based on the singular points found on the outside region of a palm. Six typical cases of singular points are defined for the palmprint classification. Experimental results show that both the line based feature extraction and matching strategies, and the singular point based method for palmprint classification are effective.

Chapter Five reports a dynamic selection scheme for offline palmprint authentication which measures global texture features and then detects local interesting points. Palmprint patterns can be well described by textures: the texture energy of a palmprint possesses a large variance between different classes while remaining high compactness within the same class. The coarse-level classification using global texture features is effective and essential for reducing the number of samples for further processing at fine level. Layered searching scheme based on the interesting points improves the system efficiency further. The use of Hausdorff distance as the matching criterion can handle the partial occluded palmprint patterns, which makes our algorithm more robust. The experimental results demonstrate that this dynamic scheme has high performance in terms of effectiveness, accuracy, robustness and efficiency.

Chapter Six presents three palmprint registration methods based on online palmprints. Two of these methods are designed to extract a fixed size square area as the central part palmprint sub-image for feature extraction, while the remaining one uses an inscribed circle as the central part of the palmprint sub-image. After that, some analysis has been performed to test the robustness of these methods. The inscribed circle method utilizes a large circle which is close to the boundary of a palm, while square based method uses a fixed size 128 × 128 area. However, larger surface area for the palmprint features extraction leads to a longer computation time. We can say that the square method has the advantage of faster processing on the feature extraction while the inscribed circle method has more features obtained. By the way, the inscribed circle method provides a means of palmprints classification which is useful in a large database. Finally, we can conclude that the selection of a palmprint segmentation method is dependent on the feature extraction algorithm and the application of the system.

Chapter Seven reports a novel textured-based feature extraction method for personal authentication that uses low resolution online palmprint images. A palmprint is considered as a texture image, so an adjusted Gabor filter is employed to capture the texture information of palmprints. Combined with the effects of preprocessing techniques and rotational robustness of the filter, the matching process is translational and rotational invariant. Experimental results illustrate the effectiveness of this method.

Chapter Eight proposes a novel approach on palm line extraction and matching. A set of directional line detectors has been devised for effective palm line extraction. Palm lines are represented by their chain code to preserve the details of its structure. Then, palmprints are matched by matching the points on their palm lines. The experiment results from a general database demonstrate the proposed approach is more powerful for palmprint verification than the 2-D Gabor algorithm when FRR < 0.91%, and the EER of our approach is also decreased from 0.6% of 2-D Gabor algorithm to 0.4%. Furthermore, another experiment designed for testing its robustness against dirty palms confirms the advantage of our approach over 2-D Gabor algorithm, where the EER of our approach is 0.63% less than that of 2-D Gabor algorithm. In addition, the average memory requirement for a palmprint is only 267 bytes and the average processing time, including preprocessing of palmprint image and matching, is only 0.6 second, which has proved the practical value of our approach.

Chapter Nine discusses two novel algebraic features called Fisherpalms and Eigenpalms. Fisher's Linear Discriminant is used to project the palmprint image from the very high dimensional original palmprint space to the very low dimensional Fisher palmprint space, in which the ratio of the determinant of the between-class scatter to that of the within-class scatter is maximized. From the Fisherpalms based palmprint recognition system, the images with resolution 32×32 are optimal for medium security biometric systems while those with resolution 64×64 are optimal for high security biometric systems. For palmprints with resolution 64×64, these accuracies are 99.05% and 99.25%, obtained in one-to-one matching test and one-to-300 matching test, respectively. For palmprints with resolution 128×128, accuracies of 99.00% and 99.20% are obtained in one-to-one matching test and one-to-300 matching test, respectively. The average testing time for the images with these resolutions in one-to-300 matching is less than 0.4s, which is fast enough for real-time palmprint recognition. Similarly, eigenpalms method is developed by using the K-L transform algorithm. It can represent the principal components of the palmprints fairly well. The features are extracted by projecting palmprint images into an eigenpalms subspace. The weighted Euclidean distance classifier is applied. A correct recognition rate of up to 99% can be obtained using eigenpalms. Finally, it can be seen that both fisherpalms and eigenpalms using the algebraic features from palmprint can achieve high recognition rates.

Chapter Ten reports a novel feature extraction method by converting a palmprint image from a spatial domain to a frequency domain. A feature extraction method using Fourier Transform is first introduced. It can be seen that the similar palmprints resemble to each other when converted to frequency domain while different palmprints are separated far away from each other. Experimental results show that the identification accuracy is 95.48%. Then, another transform-based approach using Wavelet signatures is reported. It is developed by characterizing a palmprint with a set of statistical signatures. The palmprint is transformed into the wavelet domain, and then cluster the predominant structures by context modeling according to the appearances of the principal lines in each subband. Some of the signatures are used to classify the palmprints hierarchically, and all the signatures are used to calculate the weighted distances between the palmprints and the database. Experiment is performed to classify 50 individuals into eight categories and the corrected recognition rate is 98%.

Chapter Eleven reveals a novel algorithm for palmprint classification using principal lines. Principal lines are defined and characterized by their position and thickness. A set of directional line detectors is devised for principal line extraction. Then, some rules for palmprint classification are presented. By using these detectors, the potential line initials of the principal lines are extracted. Based on these extracted potential line initials, the principal lines are extracted completely using a recursive process. The local information about the extracted part of the principal line is used to decide a ROI and then a suitable line detector is chosen to extract the next part of the principal line in this ROI. Then, some rules for palmprint classification is presented. The palmprints are classified into six categories according to the number of the principal lines and their intersections. The proposed algorithm has 96.03% accuracy rate on the palmprints classification.

Chapter Twelve reveals a new idea on the online palmprint identification based on hierarchical multi-feature coding scheme. It can help to efficient and effective palmprint verification and identification in a large database by applying a coarse-to-fine matching strategy. There are four-level features defined: global geometry based key point distance, global texture energy, fuzzy "interest" line and local directional texture energy. In contrast to the existing methods using a fixed mechanism for feature extraction and similarity measurement, this approach (using multiple features) has higher performance. Experimental results show that this approach is robust for a large palmprint database.

Chapter Thirteen introduces novel ideas for palmprint acquisition including a specially designed user interface and real-time palmprint acquisition capabilities, high quality palmprint image with reasonable price of the device are achieved. Requirement analyses on the proposed system in terms of the minimum feature size, spatial resolution and user interface are performed. During the implementation stage, we optimize the optical parameters in order to achieve the best performance in terms of palmprint image quality and size of the system. The fastest time to obtain a palmprint image at spatial resolution of 150 dpi is $0.09s$. Our proposed system is able to obtain features from a palm including principal lines, wrinkles and ridge texture. We have implemented the system at our laboratory entrance for access control since early March 2003 to demonstrate the effectiveness and viability of the system.

In Chapter Fourteen, we first present a novel biometric system based on the palmprint. It can accurately identify a person in real time and is suitable for various civil applications such as access control. Experimental results show that the proposed system can identify 400 palms with a low false acceptance rate, 0.02%, and a high genuine acceptance rate, 98.83%. For verification, the system can operate at a false acceptance rate, 0.017% and a false rejection rate, 0.86%. Experimental results including accuracy, speed and robustness demonstrate that the proposed palmprint authentication system is comparable with other hand-based biometrics systems, such as hand geometry and fingerprint verification system and is practical for real-world applications.

15.2   Future Work

A palmprint identification prototype for civil applications is demonstrated in Chapter 14. It is the world's first civilian palmprint identification system for personal identification. We employed the algorithm reported in Chapter 7 and robustness test discussed in Chapter 14, that it can handle different real-world challenges accordingly. Although this prototype is so successful, there are different research aspects for the future work, such as:

- A) Higher performance
- B) Distinctiveness of palmprint
- C) Permanence of palmprint
- D) Palmprint databases
- E) Application related issues
- F) Privacy concerns

A) Higher performance
We present a novel personal identification system based on the palmprint in Chapter 14. It can accurately identify a person in real time, which is suitable for various civil applications such as access control. Experimental results show that the proposed system can identify 400 palms with a low false acceptance rate, 0.02%, and a high genuine acceptance rate, 98.83%. Although this result is promising, we would like to develop new algorithms which can handle more users for the 1-to-many identification, with even higher performance in terms of accuracy and number of users. That is, it can achieve very low error rate for even more users, i.e. > 10,000, for the 1-to-many identification.

B) Distinctiveness of palmprint
It is not uncommon to ask a question about a biometric, whether it is unique enough for a relatively large user database. For the distinctiveness, here we interest in the information in the palmprint, whether it is sufficiently enough for identifying a person from large population. In other words, can we find out some palmprints from different palms but they are very similar? To investigate the distinctiveness of palmprints, two kinds of experiments can be conducted. In the first kind of experiment, each image in a large palmprint database compares all the others samples in this dataset. The matching score is considered as a genuine matching score if two palmprint images come from the same palm; otherwise, it is considered as an

imposter matching score. Each palm in this dataset has several samples stored. When two palms are compared, many genuine and also imposter matching scores are generated. We take the mean of all genuine matching scores of each palm. Similarly, we take the mean of all the imposter matching scores to obtain the dissimilarity of a palm to all the other palms. Therefore, each palm has a mean of genuine matching scores and a mean of imposter matching scores. When we have collected a huge palmprint database, we could perform this testing to proof the distinctiveness of palmprint.

Another experiment involves the testing of identical twins. Other biometrics such as face and DNA cannot pass this test. To conduct this test, we match a palmprint in the twin database with his/her identical twin sibling to produce imposter matching scores. Chapter 14 has reported the experiment details. From the results, we found that the identical twins' palmprint can easily be separated, just like twins' fingerprints [127]. We can observe that identical twins' palmprints still contain a lot of distinctive information.

C) Permanence of palmprint

Permanence is another important issue for biometric identification. Each biometric has some variations. Even for DNA, mutation is one of the means to change it. Face change depends on our weight, age and living styles. Undoubtedly, palmprints have similar situations. Our hands are growing from childhood to adulthood, which implies that palmprints is changing at that period of time. Our previous study shows that our method can recognize palmprints collected with a period of time over several years [65]. Figures 15-1 show three pairs of palmprint images collected with periods of 1,378 days, 1,430 days and 1,256 days, respectively. In term of image intensity, they have some difference since they are collected by different capture devices. In term of the features, principal lines and wrinkles, they are completely stable. We do not discover any observable change from those features.

D) Palmprint databases

The offline and online palmprint databases were reported in Chapter 2 already. As the advance in technology, it is much easier to get the biometrics signal online in real-time. The future work in the palmprint collection is focused on the online one, and the database collection strategies can be arranged as follow:

(a) same person with different time occasions (i.e. once a year, for consecutive 5 years) of palmprint images → to test the permanence of palmprints

(b) collect more persons' palmprint in the database → to test the distinctiveness of palmprints

(c) collect palmprints of different age group distributions → to test the robustness of palmprint in a wide range of age group distributions

(d) collect palmprints of different job natures → to test the robustness of palmprint in a wide range of job natures

We can follow these ideas to build up a large scale palmprint database so that different experiments can be performed.

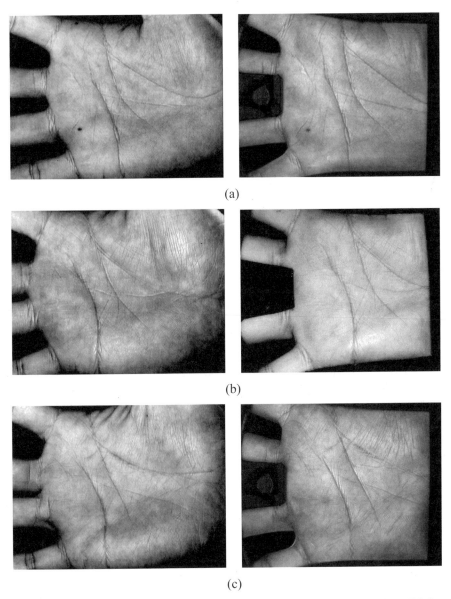

Figure 15-1. Three pairs of palmprint images with long intervals, a) 1,378 days, b) 1,430 days and c) 1,256 days

In order to promote the palmprint research, we published part of our palmprint database on the Internet [128] so that more researchers can join the palmprint researches. More algorithms targeted to solve different issues can be evolved finally. Finally, more papers can be published on different aspect of palmprint researches, including algorithm, hardware device, applications, etc.

E) *Application related issues*

Although a palmprint prototype has built successfully and being used, it is still a new technology to be accepted by the public. There is not much confidence for the public on using this technology. In order to promote the palmprint identification system, apart from the system performance perspective, we should take care on the size of the system, user acceptance, networking architecture, etc.

● *The size of the system*

Currently, the size of our system is similar to that of the 3D hand geometry system. Although it is not a problem on installing this system outside the building entrance, it is not a good idea for using it in some situation like computer logon. All in all, smaller size could make the system to be used in more possible applications.

● *User acceptance*

Palmprint is a hand-based biometric technology, exploiting the features in the inner surface of our palm for personal identification. Therefore, we expect that the palmprint identification system will receive a high user acceptance, like fingerprint, hand geometry and hand vein [6, 8, 118]. On the other hand, as it is a new biometric, not much confidence to the public and users on using it. We need to promote it via different ways like user education, public seminars, and collaboration with government departments, etc.

● *Networked architecture*

Our current palmprint identification system is a standalone system, which cannot fulfill the needs of many day to day applications, such as access control of buildings with several gates. With the expertise and experiments of our current system, we can extend the current system to a networked version so that multiple access points can be installing the system to form a networked personal identification solution. After that, it can be used for various applications such as access control and time attendance management.

F) *Privacy concerns*

Privacy concern of a biometric identifier is another important issue affecting the deployment of a particular biometrics. Like fingerprint technology, some people may be afraid of their fingerprint data being used by third parties for criminal investigation leading them reluctant to adopt it. In order to solve this problem, different aspects should be take care of. For the palmprint data, our system only stores the templates of a palmprint, but not the palmprint image itself. These palmprint templates are irreversible so that no worry on someone being taken away the templates and create another record or doing other illegal matters. Secondly, our system stores the template and general user information separately so that the application developers cannot read the templates illegally. Thirdly, we should declare that our palmprint identification system will not be used on any criminal system so that more confidence to the public can be achieved. In future, we can consider adopting some encryption methods on the palmprint data of each of the communication channels involved so that no one can take the templates out the system for the enrollment on other applications.

REFERENCES

[1]   A. Jain, R. Bolle and S. Pankanti (eds.), *Biometrics: Personal Identification in Networked Society*, Kluwer Academic Publishers, Boston Hardbound, January, 1999.

[2]   P. Fox, "Want to Save Some Money? Automate Password Resets", http://www.computerworld.com/securitytopics/security/story/0,10 801,61964,00.html.

[3]   G. S. K. Fung, R. W. H. Lau and J. N. K Liu, "A Signature Based Password Authentication Method", *Proceedings of IEEE International Conference on Systems, Man, and Cybernetics (SMC'97)*, Orlando, Florida, USA, pp. 631-636, October, 1997.

[4]   D. Zhang (Ed.), *Biometric Solutions for Authentication in an E-world*, Kluwer Academic Publishers, Boston Hardbound, July, 2002.

[5]   D. G. Joshi, Y. V. Rao, S. Kar, V. Kumar and R. Kumar, "Computer-Vision-Based Approach to Personal Identification Using Finger Crease Pattern", *Pattern Recognition*, vol. 31, no. 1, pp. 15-22, 1998.

[6]   R. Sanchez-Reillo, C. Sanchez-Avila and A. Gonzalez-Marcos, "Biometric Identification through Hand Geometry Measurements", *IEEE Transactions on Pattern Analysis and Machine Intelligence*, vol. 22, no 10, pp. 1168-1171, 2000.

[7]   R. Wildes, "Iris Recognition: an Emerging Biometric Technology", *Proceeding of the IEEE*, vol. 85, no. 9, pp. 1349-1363, 1997.

[8]   A. K. Jain, L. Hong and R. Bolle, "On-line Fingerprint Verification", *IEEE Transactions on Pattern Analysis and Machine Intelligence*, vol. 19, no 4, pp. 302-314, 1997.

[9]   B. Miller, "Vital Signs of Identity", *IEEE Spectrum*, vol. 31, no. 2, pp. 22-30, February, 1994.

[10] D. Zhang, *Automated Biometrics: Technologies and Systems*, Kluwer Academic Publishers, Boston Hardbound, May, 2000.

[11] Identix Press Releases, "Mexican Government Adopts FaceIt® Face Recognition Technology to Eliminate Duplicate Voter Registrations in Upcoming Presidential Elections", http://www.shareholder.com/identix/ ReleaseDetail.cfm?ReleaseID=53264.

[12] *Biometrics Market Report 2003-2007*, International Biometric Group, 2003.

[13] S. Pankanti, R. M. Bolle, A. Jain, "Biometrics: The Future of Identification", *IEEE Computer*, vol. 33, no. 2, pp. 46-49, 2000.

[14] M. Golfarelli, D. Maio and D. Maltoni, "On the Error-Reject Trade-Off in Biometrics Verification Systems", *IEEE Transactions on Pattern Analysis and Machine Intelligence*, vol. 19, no. 7, pp. 786-796, 1997.

[15] A. K. Jain and S. Pankanti, "Automated Fingerprint Identification and Imaging Systems", *Advances in Fingerprint Technology*, 2nd Ed. (H. C. Lee and R. E. Gaensslen), Elsevier Science, New York, 2001.

[16] A. K. Jain, A. Ross and S. Prabhakar, "An Introduction to Biometric Recognition", *IEEE Transactions on Circuits and Systems for Video Technology, Special Issue on*

Image- and Video-Based Biometrics, vol. 14, no. 1, pp. 4-20, January, 2004.

[17]   D. R. Richards, "Rules of thumb for biometric systems", *Security Management*, vol. 39, no. 10, pp. 67-71, 1995.

[18]   D. Zhang and W. Shu, "Two Novel Characteristics in Palmprint Verification: Datum Point Invariance and Line Feature Matching", *Pattern Recognition*, vol. 32, no. 4, pp. 691-702, 1999.

[19]   L. Coetzee, and E. C. Botha, "Fingerprint recognition in low quality images", *Pattern Recognition*, vol. 26, no. 10, pp. 1441-1460, 1993.

[20]   D. McGuire, "Virginia Beach Installs Face-Recognition Cameras", *The Washington Post*, 2002, http://www.washingtonpost. com/ac2/wp-dyn/A19946-2002Jul3.

[21]   International Biometric Group's Consumer Response to Biometrics, 2002, http://www.ibgweb.com/reports/public/reports/ facial-scan_perceptions.html.

[22]   N. Ratha, S. Chen, K. Karu and A. K. Jain, "A Real-Time Matching System for Large Fingerprint Databases", *IEEE Transactions on Pattern Analysis and Machine Intelligence*, vol. 18, no. 8, pp. 799-813, 1996.

[23]   K. Karu and A. K. Jain, "Fingerprint Classification", *Pattern Recognition*, vol. 29, no. 3, pp. 389-404, 1996.

[24]   R. Cappelli, A. Lumini, D. Maio and D. Maltoni, "Fingerprint Classification by Directional Image Partitioning", *IEEE Transactions on Pattern Analysis and Machine Intelligence*, vol. 21, no. 5, pp. 402-421, 1999.

[25]   D. Maio and D. Maltoni, "Direct Gray-Scale Minutiae Detection in Fingerprints", *IEEE Transactions on Pattern Analysis and Machine Intelligence*, vol. 19, no. 1, pp. 27-40, 1997.

[26]   J. Berry, "The history and development of fingerprinting", *Advances in Fingerprint Technology*, (H.C. Lee and R.E. Gaensslen, ed.s), CRC Press, Florida, pp. 1-39, 1994.

[27]   INSPASS Project, http://www.panynj.gov/aviation/inspass main.htm.

[28]   R. Sanchez-Reillo and C. Sanchez-Acila, "Access Control System with Hand Geometry Verification and Smart Cards", *IEEE Aerospace and Electronics Systems Magazine*, vol. 15, no. 2, pp. 45-48, 2000.

[29]   Fujitsu Laboratories Limited, "Biometric Mouse with Palm Vein Pattern Recognition Technology", http://pr.fujitsu.com/en/news/2002/08/28.html.

[30]   Fujitsu Laboratories Limited, "Contactless Palm Vein Pattern Biometric Authentication System", http://pr.fujitsu.com/en/news/2003/03/31.html.

[31]   W. Shu and D. Zhang, "Automated Personal Identification by Palmprint", *Optical Engineering*, vol. 37, no. 8, pp. 2659-2362, 1998.

[32]   M. Elecion, "Automatic fingerprint identification", *IEEE Spectrum*, vol. 10, no. 9, pp. 36-45, 1973.

[33]   J. Wayman, A. K. Jain, D. Maltoni, and D. Maio, *Biometric Systems: Technology, Design and Performance Evaluation*, Springer-Verlag, UK, 2003.

[34]   Thoe Pavlidis, *Algorithms for Graphics and Image Processing*, Rockville, Md.: Computer Science Press, Inc., 1982.

[35]   V. S. Srinivasan, N. N. Murthy, "Detection of Singular Points in Fingerprint Images", *Pattern Recognition*, vol. 25, no. 2, pp. 139-153, 1992.

[36]   C. L. Wilson, G. T. Candela and C. I. Watson, "Neural Network Fingerprint Classification", *Journal of Artificial Neural Networks*, vol. 1, no. 2, pp. 1-25, 1993.

[37]   Paul S. Wu and Ming Li, "Pyramid edge detection based on stack filter", *Pattern Recognition Letters*, vol. 18, no. 4, pp. 239-248, 1997.

[38]   T. S. Chan and K. K. Yip, "Line detection algorithm", *Proceeding of 13th ICPR*, pp. 126-130, 1996.

[39]   J. B. Burns, A. R. Hanson and E. M. Riseman, "Extracting straight lines", *IEEE Transactions on Pattern Analysis and Machine Intelligence*, vol. 8, no. 4, pp. 425-455, 1986.

[40]   Azriel Rosenfeld and Avinash C. Kak, *Digital Picture Processing*, London: Academic Press, 1982.

[41]   J. F. Keegan, "How can you tell if two line drawings are the same?", *Computer Graphics and Image Processing*, vol. 6, no. 1, pp. 90-92, 1977.

[42]   Grolier Incorporated, *The Encyclopedia American*, Grolier, USA, 1995.

[43]   P. Baltscheffsky and P. Anderson, "The Palmprint Project: Automatic Identity Verification by Hand Geometry", *Proc. 1986 International Carnahan Conference on Security Technology*, pp. 229-234, 1986.

[44]   W. Shu and D. Zhang, "Palmprint Verification: An Implementation of Biometric Technology", *Proc. 14th International Conference on Pattern Recognition*, pp. 219-221, 1998.

[45]   K. I. Laws, "Textured Image Segmentation", *Ph.D Thesis*, University of Southern California, 1980.

[46]   G. Borgefors, "Hierarchical Chamfer Matching: A Parametric Edge Matching Algorithm", *IEEE Transactions on Pattern Analysis and Machine Intelligence*, vol. 10, pp. 849-865, 1988.

[47]   D. P. Huttenlocher, G. A. Klanderman and W. J. Rucklidge, "Comparing Images Using the Hausdorff Distance", *IEEE Transactions on Pattern Analysis and Machine Intelligence*, vol. 15, pp. 850-863, 1993.

[48]   J. You, E. Pissaloux, J. L. Hellec and P. Bonnin, "A Guided Image Matching Approach Using Hausdorff Distance with Interesting Points Detection", *Proc. of IEEE International Conference on Image Processing*, pp. 968-972, 1994.

[49]   W. Li, "Authenticating Personal Identities Using Palmprint Recognition", *Ph.D. Thesis*, The Hong Kong Polytechnic University, 2004.

[50]   W. K. Kong, D. Zhang, W. Li, "Palmprint feature extraction using 2-D Gabor Filters, *Pattern Recognition*, vol. 36, pp. 2339-2347, 2003.

[51]   D. Gabor, "Theory of communications", *Journal of IEE*, vol. 93, pp. 429-457, 1946.

[52]   C. K. Chui, *An Introduction to Wavelets*, Academic Press, Boston, 1992.

[53]   J. G. Daugman, "Two-dimensional spectral analysis of cortical receptive field profiles", *Vision Research*, vol. 20, pp. 847-856, 1980.

[54]   J. Daugman, "Uncertainty relation for resolution in space, spatial frequency and orientation optimized by two-dimensional visual cortical filters", *Journal of the Optical Society of America A*, vol. 2, pp. 1160-1169, 1985.

[55]   A. Jain and G. Healey, "A multiscale representation including opponent color features for texture recognition", *IEEE Transactions on Image Processing*, vol. 7, no. 1, pp. 124-128, 1998.

[56]   D. Dunn and W. E. Higgins, "Optimal Gabor filters for texture segmentation", *IEEE Transactions on Image Processing*, vol. 4, no. 4, pp. 947-964, 1995.

[57]   A. C. Bovik, M. Clark and W. S. Geisler, "Multichannel texture analysis using localized spatial filters", *IEEE Transactions on Pattern Analysis and Machine Intelligence*, vol. 12, no. 1, pp. 55-73, 1990.

[58]   J. Daugman, "High confidence visual recognition of persons by a test of statistical independence", *IEEE Transactions on Pattern Analysis and Machine Intelligence*, vol. 15, no. 11, pp. 1148-1161, 1993.

[59]   A. K. Jain, S. Prabhakar, L. Hong and S. Pankanti, "Filterbank-based fingerprint matching", *IEEE Transactions on Image Processing*, vol. 9, no. 5, pp. 846-859, 2000.

[60]   L. Hong, Y. Wan and A. Jain, "Fingerprint image enhancement algorithm and performance evaluation", *IEEE Transactions on Pattern Analysis and Machine Intelligence*, vol. 20, no. 8, pp. 777-789, 1998.

[61]   C. J. Lee and S. D. Wang, "Fingerprint feature extraction using Gabor filters", *Electronic Letters*, vol. 35, no. 4, pp. 288-290, 1999.

[62]   M. J. Lyons, J. Budynek and S. Akamatsu, "Automatic classification of single facial images", *IEEE Transactions on Pattern Analysis and Machine Intelligence*, vol. 21, no. 12, pp. 1357-1362, 1999.

[63]   B. Duc, S. Fischer and J. Bigun, "Face authentication with Gabor information on deformable graphs", *IEEE Transactions on Image Processing*, vol. 8, no. 4, pp. 504-516, 1999.

[64]   Y. Adini, Y. Moses and S. Ullman, "Face recognition: The problem of compensation for changes in illumination direction", *IEEE Transactions on Pattern Analysis and Machine Intelligence*, vol. 19, no. 7, pp. 721-732, 1997.

[65]   D. Zhang, W. Kong, J. You and M. Wong, "Online palmprint identification", *IEEE Transactions on Pattern Analysis and Machine Intelligence*, vol. 25, no. 9, pp. 1041-1050, 2003.

[66]   C. C Han, H. L. Chen, C. L. Lin and K. C. Fan, "Personal authentication using palm-print features", *Pattern Recognition*, vol. 36, no. 2, pp. 371-381, 2003.

[67]   A. Kumar, D. C. M. Wong, H. C. Shen, A. Jain, "Personal Verification using Palmprint and Hand Geometry Biometric", *Lecture Notes in Computer Science*, vol. 2688, pp. 668-678, 2003.

[68]   R. M. Haralick, "Ridges and Valleys on Digital Images", *Computer Vision, Graphics, and Image Processing,* vol. 22, pp. 28-38, 1983.

[69]   K. Liang, T. Tjahjadi and Y. Yang, "Roof edge detection using regularized cubic B-spline fitting", *Pattern Recognition*, vol. 30, no. 5, pp. 719-728, 1997.

[70]   K. R. Castleman, *Digital Image Processing*, Prentice Hall, Inc. 1996.

[71]   J. Canny, "A computational approach to edge detection", *IEEE Transactions on Pattern Analysis and Machine Intelligence*, vol. 8, no. 6, pp. 679-698, 1986.

[72]   J. R. Parker, *Algorithms for Image Processing and Computer Vision*, John Wiley & Sons, Inc., 1997.

[73]   P. N. Belhumeur, J. P. Hespanha, D. J. Kriegman, "Eigenfaces vs. Fisherfaces: recognition using class specific linear projection", *IEEE Transactions on Pattern Analysis and Machine Intelligence*, vol. 19, no. 7, pp. 711-720, 1997.

[74]   S. Bengio, J. Mariethoz, S. Maroel, "Evaluation of Biometric Technology On XM2VTS", *IDIAP Research Report 01-21*, Dalle Molle Institute for Perceptual Artificial Intelligence, 2001.

[75]   Y. Cui, D. Swets, J. Weng, "Learning-based hand sign recognition using SHOSLIF-M", *International Conference on Computer Vision*, pp. 631-636, 1995.

[76]   R. O. Duda, P. E. Hart, D. G. Stork, *Pattern Classification,* John Wiley & Sons, Inc., 2001.

[77]   N. Duta, A. K. Jain, K. V. Mardia, "Matching of Palmprints", *Pattern Recognition Letters*, vol. 23, no. 4, pp. 477-485, 2001.

[78]   R. A. Fisher, "The use of multiple measures in taxonomic problems", *Annals of Eugenics*, vol. 7, pp. 179-188, 1936.

[79]   L. D. Harmon, "The recognition of faces", *Scientific American*, vol. 29, pp. 71-82, 1973.

[80]   W. Li, D. Zhang, Z. Xu, "Palmprint identification by Fourier Transform", *International Journal of Pattern Recognition and Artificial Intelligence*, vol. 16, no. 4, pp. 417-432, 2002.

[81]   K. Liu, Y. Cheng, J. Yang, "Algebraic feature extraction for image recognition based on an optimal discriminant criterion", *Pattern Recognition*, vol. 26, no. 6, pp. 903-911, 1993.

[82]   J. You, W. Li, D. Zhang, "Hierarchical palmprint identification via multiple feature extraction", *Pattern Recognition*, vol. 35, no. 4, pp. 847-859, 2002.

[83]   P. C. Yuela, D. Q. Dai, G. C. Feng, "Wavelet-based PCA for human face recognition", *IEEE Southwest Symposium on Image Analysis and Interpretation*, pp. 223-228, 1998.

[84]   H. Peng, D. Zhang, "Dual eigenspace method for human face recognition", *IEE Electronics Letter*, vol. 33, no. 4, pp. 283-284, 1997.

[85]   M. Turk, A. Pentland, "Eigenfaces for recognition", *Journal of Cognitive Neuroscience*, vol. 3, no. 1, pp. 71-86, 1991.

[86]   Y. Zhu, Y. Wang and T. Tan, "Biometric Personal Identification Based on Handwriting", *Proc. 15th International Conference on Pattern Recognition (ICPR)*, vol. 2, pp.801-804, Barcelona, Spain, September, 2000.

[87]   E. C. Titchmarsh, *Introduction to the Theory of Fourier Integral,* Oxford University Press, New York, 1948.

[88]   J. W. Cooley, P. A. W. Lewis and P. D. Welch, "Historical notes on the fast fourier transform", *IEEE Transactions on Audio and Electroacoustics*, vol. 15, no. 2, pp. 76-79, 1967.

[89]   J. W. Cooley, P. A. W. Lewis and P. D. Welch, "Application of the fast Fourier transform to computation of Fourier integrals, Fourier series, and convolution integrals", *IEEE Transactions on Audio and Electroacoustics*, vol. 15, no. 2, pp. 79-84, 1967.

[90]   E. O. Brigham, *The Fast Fourier Transform*, Prentice-Hall, Englewood Cliffs, N.J., 1974.

[91]   H. C. Andrews, *Computer Techniques in Image Processing*, Academic Press, New York, 1970.

[92]   R. G. Gonzalez and R. E. Woods, *Digital Image Processing*, Addison-Wesley Publishing Company, 1993.

[93]   C. Liu and H. Wechsler, "A shape- and texture-based enhanced Fisher classifier for face recognition", *IEEE Transactions on Image Processing*, vol. 10, pp. 598-608, April, 2001.

[94]   I. Daubechies, *Ten lectures on wavelets*, Philadelphia, PA: SIAM, 1992.

[95]   S. Mallat, "A theory for multiresolution signal decomposition: The wavelet representation", *IEEE Transactions on Pattern Analysis and Machine Intelligence*, vol. 11, pp. 674-693, July, 1989.

[96]   S. Mallat and S. Zhong, "Characterization of signals from multiscale edges", *IEEE Transactions on Pattern Analysis and Machine Intelligence*, vol. 14, pp. 710-732, July, 1992.

[97]   M. Vetterli and C. Herley, "Wavelet and filter banks: theory and design", *IEEE Transactions Signal Processing*, vol. 40, pp. 2207-2232, September, 1992.

[98]   S. G. Chang, B. Yu and M. Vetterli, "Spatially adaptive wavelet thresholding with context modeling for image denoising", *IEEE Transactions on Image Processing*, vol. 9, pp. 1522-1531, September, 2000.

[99]   X. Wu, "Lossless Compression of continuous-tone images via context selection, quantization, and modeling", *IEEE Transactions on Image Processing*, vol. 6, pp. 656-664, May, 1997.

[100]   Y. Yoo, A. Ortega and B. Yu, "Image subband coding using context-based classification and adaptive quantization", *IEEE Transactions on Image Processing*, vol. 8, pp. 1702-1715, December, 1999.

[101]   J. Liu and P. Moulin, "Information-theoretic analysis of interscale and intrascale dependencies between image wavelet coefficients", *IEEE Transactions on Image Processing*, vol. 10, pp. 1647-1658, November, 2001.

[102]   R. R. Coifman and D. L. Donoho, "Translation-invariant de-noising", in *Wavelet and Statistics*, A. Antoniadis and G. Oppenheim, Eds. Berlin, Germany: Springer-Verlag, 1995.

[103]   J. M. Shapiro, "Embedded image coding using zerotrees of wavelet coefficients", *IEEE Transactions Signal Processing*, vol. 41, pp. 3445-3462, December, 1993.

[104]   A. K. Jain, R. P. W. Duin and J. Mao, "Statistical pattern recognition: a review", *IEEE Transactions Pattern Analysis and Machine Intelligence*, vol. 22, pp. 4-37, January, 2000.

[105]   T. Chang and C. C. J. Kuo, "Texture analysis and classification with tree-structured

wavelet transform", *IEEE Transactions on Image Processing*, vol. 2, pp. 429-441, October, 1993.

[106]   S. Pittner and S. V. Kamarthi, "Feature extration from wavelet coefficients for pattern recognition tasks", *IEEE Transactions on Pattern Analysis and Machine Intelligence*, vol. 21, pp. 83-88, January, 1999.

[107]   A. Laine and J. Fan, "Texture classification by wavelet packet signitures", *IEEE Transactions on Pattern Analysis and Machine Intelligence*, vol. 15, pp.1186-1191, November, 1993.

[108]   M. Unser, "Texture classification and segmentation using wavelet frames", *IEEE Transactions on Image Processing*, vol. 4, pp. 1549-1560, November, 1995.

[109]   G. V. de Wouwer, P. Scheunders and D. V. Dyck, "Statistical texture characterization from discrete wavelet representation", *IEEE Transactions on Image Processing*, vol. 8, pp. 592-598, April, 1999.

[110]   W. Shu, G. Rong and Z. Bian, "Automatic palmprint verification", *International Journal of Image and Graphics*, vol. 1, no. 1, pp. 135-151, 2001.

[111]   Identix Incorporated, "TouchPrint™ PRO Full Hand Scanner", http://www.identix.com/products/pro_livescan_TPpro.html.

[112]   NEC Solutions (America), Inc., "Automated Palmprint Identification System", http://www.necsolutions-am.com/idsolutions/products/palmprint_product.cfm.

[113]   C. E. Shannon, "Communication in the presence of noise", *Proc. Institute of Radio Engineers*, vol. 37, no.1, pp. 10-21, January, 1949.

[114]   M. W. Burke, *Handbook of Machine Vision Engineering*, Chapman & Hall, Great Britian, 1996.

[115]   G. C. Holst, *CCD Arrays, Cameras, and Displays*, JCD Publishing and SPIE Optical Engineering Press, USA, 1998.

[116]   International Biometrics Group, http://www.biometricgroup.com.

[117]   J. Daugman, "High confidence personal identification by rapid video analysis of iris texture", *in Proceeding of the IEEE International Carnahan Conference on Security Technology*, pp. 50-60, 1992.

[118]   S. K. Im, H. M. Park, Y. W. Kim, S. C. Han, S. W. Kim and C. H. Kang, "An biometric identification system by extracting hand vein patterns", *Journal of the Korean Physical Society*, vol. 38, no. 3, pp. 268-272, 2001.

[119]   G. Lu, D. Zhang, and K. Wang, "Palmprint Recognition Using Eigenpalms Features", *Pattern Recognition Letters*, vol. 24, pp. 1473-1477, 2003.

[120]   D. Maio, D. Maltoni R. Cappelli, J. L. Wayman and A. K. Jain, "FVC 2000: Fingerprint verification competition", *IEEE Transactions on Pattern Analysis and Machine Intelligence*, vol. 24, no. 3, pp. 402-412, 2002.

[121]   Y. He, J. Tian, X. Luo, T. Zhang, "Image enhancement and minutiae matching in fingerprint verification", *Pattern Recognition Letters*, vol. 24, pp. 1349-1360, 2003.

[122]   C. Quek and R. W. Zhou, "Antiforgery: a novel pseudo-outer product based fuzzy neural network driven signature verification system", *Pattern Recognition Letters*, vol. 23, pp. 1795-1816, 2002.

[123]   M. J. Er, S. Wu, J. Lu and H. L. Toh, "Face recognition with radial basis function (RBF) neural networks", *IEEE Transactions on neural networks*, vol. 13, no. 3, pp. 697-710, 2002.

[124]   S. Lawrence, C. L. Giles, A. C. Tsoi and A. D. Back, "Face recognition: a convolutional neural-network approach", *IEEE Transactions on neural networks*, vol. 8, no. 1, pp. 98-113, 1997.

[125]   A. Tefas, C. Kotropoulos and L. Pitas, "Using support vector machines to enhance the performance of elastic graph matching for frontal face authentication", *IEEE Transactions on Pattern Analysis and Machine Intelligence*, vol. 23, no. 7, pp. 735-746, 2001.

[126]   T. Mansfield, G. Kelly, D. Chandler, and J. Kane, *Biometrics Product Testing Final Report*, Computing, National Physical Laboratory, Crown Copyright, U.K., (http://www.cesg.gov.uk/site/ast), 2001.

[127]   A. K. Jain, S. Prabhakar and S. Pankanti, "On the similarity of identical twin fingerprints", *Pattern Recognition*, vol. 35, no. 11, pp. 2653-2662, 2002.

[128]   Biometric Research Center, *The Hong Kong Polytechnic University*, http://www.comp.polyu.edu.hk/~biometrics.

[129]   L. Li, D. Tian, and C. Jiao, *Details of Holographic Medicine*, Chinese Medicine Technology Publisher, 2000.

[130]   W. Shen, M. Surette and R. Khanna, "Evaluation of Automated Biometrics-Based Identification and Verification Systems", *Proc. IEEE Special Issue on Automated Biometrics,* vol. 85, no. 9, pp. 1464-1478, 1997.

[131]   D. W. Fawcett, *Bloom & Fawcett: Concise Histology*, Chapman and Hall, International Thomson Publishing, New York, 1997.

[132]   J. Napier, *Hands*, George Allen & Unwin, London, 1980.

[133]   D. Maltoni, D. Maio, A. K. Jain, and S. Prabhakar, *Handbook of Fingerprint Recognition*, Springer Verlag, June, 2003.

[134]   A. Kumar and H. C. Shen, "Recognition of Palmprint using Eigenpalms", Proc.CVPRIP-2003, Cary (North Carolina), 2003.

[135]   A. Kumar and H. Shen, "Recognition of palmprints using wavelet-based features", *Proc. Intl. Conf. Sys., Cybern., SCI-2002*, Orlando, Florida, July, 2002.

[136]   X. Wu, D. Zhang and K. Wang, "Fisherpalms based Palmprint recognition", *Pattern Recognition Letters*, vol. 24, no. 15, pp. 2829-2838, 2003.

[137]   W. Li, D. Zhang and Z. Xu, 2003, "Image alignment based on invariant features for palmprint identification", *Signal Processing: image Communication*, vol. 18, no. 5, pp. 373-379, 2003.

[138]   Q. Dai, Y. Yu and D. Zhang, "A palmprint classification method based on structure features", *Pattern Recognition and Artificial Intelligence*, vol. 15, no. 3, pp. 112-116, 2002.

[139]   K. K. Benke, D. R. Skinner and C. J. Woodruff, "Convolution operators as a basis for objective correlates for texture perception", *IEEE Transactions on SMC*, vol. 18, pp. 158-163, 1988.

[140] J. You, W. Kong, D. Zhang and K. Cheung, "On Hierarchical Palmprint Coding with Multi-Features for Personal Identification in Large Databases", *IEEE Transactions on Circuits and Systems for Video Technology*, vol. 14, no. 2, pp. 234-243, 2004.

[141] R. M. Haralick, "Statistical and structural approaches to texture", *Proc. IEEE*, vol. 67, pp. 786-804, 1979.

[142] J. You and P. Bhattacharya, "A Wavelet-based coarse-to-fine image matching scheme in a parallel virtual machine environment", *IEEE Transactions on Image Processing*, vol. 9, no. 9, pp. 1547-1559, 2000.

[143] J. You and H. A. Cohen, "Classification and segmentation of rotated and scaled textured images using texture 'tuned' masks", *Pattern Recognition*, vol. 26, pp. 245-258, 1993.

INDEX

Palmprint Authentication

This is the first book to systematically provide a comprehensive introduction to palmprint technologies and explore how to design the corresponding system. It reveals automatic techniques for palmprint authentication, from the approach based on offline palmprint images, to the current state-of-the-art algorithm using online palmprint images. It is suitable for different levels of readers: those who want to learn more about palmprint technology, and those who wish to understand, participate, and/or develop a palmprint authentication system. The book can be used as a text book or reference for graduate or senior undergraduate courses on the subject. It can also be used by researchers in the corresponding fields.

David Zhang graduated in Computer Science from Peking University in 1974. He received his MSc and PhD in Computer Science and Engineering from the Harbin Institute of Technology (HIT) in 1983 and in 1985, respectively. From 1986 to 1988 he was a Postdoctoral Fellow at Tsinghua University and then an Associate Professor at the Academia Sinica, Beijing. In 1994 he received his second PhD in Electrical and Computer Engineering from the University of Waterloo, Ontario, Canada. Professor Zhang is currently at The Hong Kong Polytechnic University where he is the Founding Director of the Biometrics Research Centre (UGC/CRC) (www.comp.polyu.edu.hk/~biometrics/) supported by the Hong Kong SAR Government. He also serves as Adjunct Professor in Tsinghua University, Shanghai Jiao Tong University, HIT, and the University of Waterloo. Professor Zhang's research interests include automated biometrics-based authentication, pattern recognition, and biometric technology and systems. He is the founder and Editor-in-Chief, International Journal of Image and Graphics (IJIG) (www.worldscinet.com/ijig/ijig.shtml); Book Editor, Kluwer International Series on Biometrics (KISB) (www.wkap.nl/prod/s/KISB); Program Chair, the First International Conference on Biometrics Authentication (ICBA), and Associate Editor of more than ten international journals including IEEE Trans on SMC-A/SMC-C, Pattern Recognition, etc. He is the author of more than 130 journal papers, twenty book chapters and nine books. As a principal investigator, he has since 1980 brought to fruition many biometrics projects and won numerous prizes. Recently, his Palmprint Identification System won a Silver Medal at the Seoul International Invention Fair, a Special Gold Award, a Gold Medal, and a Hong Kong Industry Award. Professor Zhang holds a number of patents in both the USA and China and is a current Croucher Senior Research Fellow.